Telecommunications Policy and Management

J.M.Harper

Pinter Publishers
London and New York

© J. M. Harper, 1989

First published in Great Britain in 1989 by
Pinter Publishers Limited,
25 Floral Street, London WC2E 9DS

British Library Cataloguing in Publication Data

A CIP catalogue record for this book is available from the British Library

Library of Congress Cataloging-in-Publication Data

Harper, J. M. (John M.)
    Telecommunications policy and management / J.M. Harper.
        p.   cm.
    Includes bibliographical references.
    ISBN 0-86187-798-5
    1. Telecommunication policy—Developing countries.
    2. Telecommunication—Developing countries—Management.
    3. Telecommunication systems—Developing countries. I. Title.
HE8635.H47   1989
384'.068—dc20   89-39048
                                                        CIP

Typeset by Saxon Ltd., Derby.
Printed and bound in Great Britain by Biddles Ltd.

Contents

Foreword
Sir Donald Maitland   xi

Preface and acknowledgements   xiii
Dedication   xv
1   Introduction   1

Part I Policy and management   5

2   Constitutional issues   7
        Constitutional developments   7
        The position in Britain   7
        Implications for developing countries   9
        Separation of posts and telecommunications   10
        Competition in telecommunications   11
        Conclusions   13

3   Organization   15
        Overall structures   15
        Tiers of organization   21
        The quality function   22
        Network organization   22
        'Long lines' organization   24
        Organization within field units   25
        Buildings and accommodation   26
        Directories and directory enquiry   27
      ✗ New products and services   27
        International telecommunications   28
        Conclusions   29
        Summary   30

4   Accounting and control systems   33
        Telecommunications accounting systems   33
        Statistical returns and controls   35
        Subordinate unit controls   38
        Monitoring of performance   38
        Summary   39

5 Plans and planning   40
    Planning documents   41
    Forecasts   41
    Objectives   42
    Compilation   43
    The one-year plan   43
    Subordinate unit plans   44
    Answerability of field formations   45
    The five-year or medium-term plan   45
    Submission and uses of the MTP   46
    Sector plans   47
    Summary   48

✗ 6 Charging and pricing   49
    Telephone charging structures   49
    Pricing considerations   50
    Demand management and the connection charge   52
    Future developments in charging   53
    Presentation of call charges on bills   57
    Summary   57

✗ 7 Priorities and internally generated capital   58
    Priorities   58
    Determination of investment requirements   60
    Internally generated capital   62
    Externally generated funds   64
    Summary   65

✗ 8 Staff and staffing   66
    Determination of staffing requirements   66
    Requirements for basic engineering staff   67
    Requirements for operating staff   68
    Effect of automatization   68
    Requirements for office staff   69
    Managers and professional staff   70
    Recruitment   70
    Training   70
    Staff productiveness   71
    Control of overtime   73
    Measurement of productiveness   73
    Staff rewards   74
    Work-force organization   75
    General staff management considerations   77
    Industrial relations   78
    Summary   78

Part II The technology and its products 79

9 The technology in outline 81
Computing technology 81
Digital and analogue principles 83
Hardware 85
Mainframe computers 86
Minicomputers 86
Microprocessors and microcomputers 86
Supercomputers 87
Artificial intelligence 87
Software 88
Programs and programming 88
Telecommunications plant and equipment 89
Earlier techniques 89
Digital telecommunications techniques 90

10 Networks 93
Public Switched Networks (PSTNs) 93
Telex network 93
Data transmission: packet-switching 94
Integral Digital Networks (IDNs) 95
Integrated Services Digital Network (ISDN) 96
WANs and MANs 100
Changes in network basics 100
The 'intelligent' network 101
The Universal Intelligent Communications Network (UICN) 102
Cable television techniques 102
Private networks 104
Summary 106

11 The role of radio 108
Spectrum problems 108
Short-wave frequencies 109
Microwave frequencies 109
Satellite communication techniques 110
Millimetric frequencies 111
Mobile communications 111
Public mobile services linked to the public telephone network 112
Cellular mobile radio systems 112
Cordless customer premise equipment 113
Radio-paging 115
Micro-cellular techniques 115
Policy issues: rural and low-density distribution systems 115
Wider merits of cable and radio in high-density areas 116
Private systems 118
Summary 119

12 Planning, procurement and management of network plant   120
      Network planning   120
      Technical choice and replacement policy   120
      Dimensioning and routing   122
      Forecasts used in plant planning   124
      Influence of different factors   126
      Local line planning   128
      Building planning   128
      Short-term expedients   129
      Procurement of major network plant   129
      Technical authority   131
      Summary   131

13 International telecommunications and the ITU   132
      The International Telecommunications Union   132
      Plenipotentiary Conferences   133
      Administrative Conferences   133
      Administrative Council   133
      The International Frequency Registration Board (IFRB)   134
      Technical cooperation   135
      Centre for Telecommunications Development   136
      Communication between countries   136
      Intercontinental transmission   137
      Submarine telephone cables   137
      Communications satellites   138
      Optical-fibre submarine cables   138
      International telephone exchanges   139
      Summary   140

14 Customer Premise Equipment   142
      The equipment and its provision   142
      Telephones and ancillary devices   142
      Plugs and sockets   143
      Document terminals   144
      PCs with communications facilities   144
      Terminal combinations   145
      Private Manual Branch Exchanges and Call-Connect Systems   145
      Private Automatic Branch Exchanges   145
      Computer networking   146
      Local area networks   146
      Payphones   147
      Management considerations   147
      Competition in CPE and divestment of CPE work   149
      Conclusions on management and policy   151
      Summary   152

15 Network services   153
      Value-added Network Services (VANs)   153
      Database services   154
      Deferred transmission services   154
      Protocol and code conversion services: the standards problem   154
      Videotex services and telematics   156
      Sector applications: financial services and EFTPOS   157
      Policy issues: the mass use of telematics   160
      Conclusions on policy for telematics   161
      Visual services   162
      Video conferencing   163
      Vision phone   164
      Telecontrol, telemetering and alarms   164
      Summary   165

Part III Conclusions   167

16 Conclusions   169
      Principal conclusions   169

Appendices

1 Extract from the Report of the Maitland Commission   174

2 BT internal management controls, 1987   179

3 Statistical headings in the ITU *Yearbook
     of Telecommunications Statistics*   180

4 British Telecommunications statistics   183

5 Evolution of the ISDN   187

Bibliography   194

Glossary   198

Foreword

It was evident to the Independent Commission for World-Wide Telecommunications Development that most developing countries stand in need of advice on a wide range of telecommunications issues.

In its Report, *The Missing Link*, published in January 1985, the Commission analysed the problems facing telecommunications administrations in developing countries and recommended ways in which their networks might be improved and expanded.

In February 1986 the International Telecommunications Union published a study of the role of telecommunications in socio-economic development. This work, *Information, Telecommunications and Development*, supplemented an earlier study undertaken by the Union in collaboration with the Organization for Economic Cooperation and Development and published in 1983 under the title *Telecommunications for Development*. In July 1986 the Union published *Investing in Telecommunications*, which set out the considerations developing country administrations might have in mind when assessing the place of telecommunications among their investment priorities.

Each of these documents offers valuable guidance. But in the end the effectiveness of any telecommunications system depends on the way in which it is organized and managed on the ground. As a former Managing Director of the Inland Division of British Telecom, John Harper has extensive practical experience in this field; he also has the talent to impart this wisdom to others in clear terms. This was borne out by the quality of the advice he offered to the Independent Commission. In drafting Chapter 5 of *The Missing Link*, which deals with internal organization and management of telecommunications, the Commission drew heavily on his knowledge.

John Harper expressed the view then that telecommunications administrations in developing countries deserved a more detailed exposition of what effective organization and management entail. And so he set to work. This comprehensive book, in which he has distilled the benefits of his experience, is the result. I have no doubt that it will prove of great value to those who carry responsibility in this field; and not least because it contains, in Chapter 15, a lucid description of the value-added network services which are bound to make an increasingly important contribution to economic and social progress in developing countries.

Apart from its intrinsic merit, I know that this book reflects John Harper's deep personal conviction that better communications mean better lives.

Donald Maitland
June 1989

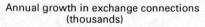

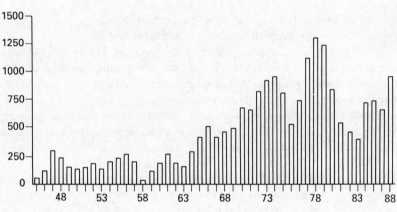

Source: BT

Preface and acknowledgements

British Telecom (BT) and its predecessor British Post Office Telecommunications (BPOT) have a long tradition of assistance to developing countries. My aim in writing this book has been to follow this tradition.

No public utility manager that I have met has ever felt that his operation was perfect, or anywhere near it. Certainly BPOT was not, nor was BT when I left it. In any case, there are no unique answers to the problems of telecommunications policy and management. The advanced countries certainly do not have a monopoly of good ideas. Each administration round the world has its own set of problems and solutions. The object of the book is not to set out a series of textbook answers, but to enable developing countries to learn from British experience — mistakes as well as successes. I should stress that the views and conclusions are my own, and not those either of BT or of NEC (UK) Ltd, my present employer.

Telecommunications policy and management are never static. Many of the matters dealt with in this book are the subject of active debate as I write. In such cases my aim has been to make the reader aware of facts and arguments and to leave him to make up his own mind.

The main body of the book is intended as an objective summary of our experience in BT, to which many people besides myself have contributed. In Chapter 7 I have tried to derive a set of suggested priorities for developing country administrations to consider in making their plans. In this Preface I would like to draw on my own experience to make some more personal suggestions about the subjects to which an administration which has problems should give special attention if it wants to improve its operations.

I believe the first step must be to set in place a modern framework of organization, accounting and statistical controls, mechanisms of answerability and planning procedures, on the lines advocated in Chapters 4 and 5. Second, I would set in hand a programme to recruit and train a sufficient cadre of good senior managers and professional engineers, if they were not already in position. Third, I would give particular attention to the calibre and training of first line supervisors, who hold the key to the efficiency of day-to-day operations. Fourth, I would see what could be done to improve service to existing and new customers, including customers in outlying areas, with the present plant and work-force. In my experience good leadership and methodical management can usually be relied on to improve the performance of one's existing assets, human or technological. Only then would I turn to the expansion, renewal and exploitation of network and other technology.

This book could not have been written without the help of many people besides myself. I owe a great debt to BT and many of its serving senior staff. Among retired BT people, I should like to thank specially Mr J. F. P. Thomas, who read and commented on the whole text. I also had invaluable help with particular chapters from Messrs G. J. Pocock and B. T. Wherry. I am also grateful to Mr A. Embedoklis, Head of the ITU Technical Cooperation Department and Herr Westendoerp, Director of the Centre For Telecommunications Development, for their advice and for providing the material about the ITU and the Centre on pages 132-136. I should like to thank BT for providing the material for the graphs in Appendix 4, the graphs on page 61 and the photographs throughout the book. I should like to thank Campus Verlag, Frankfurt-on-Main for allowing me to publish the table on page 158 and MIT Press for allowing me to publish the extract on page 160.

Finally, I am particularly grateful to Sir Donald Maitland for his help and encouragement, for allowing me to quote various passages from the Report of the Independent Commission for World-Wide Telecommunications Development (the Maitland Report) and for writing the Foreword. The book would not have been written without him.

One other debt should be acknowledged. In common with administrations in many other countries and for many years BT has learnt in all sorts of ways from the Bell System in the United States and from its successor companies. A great deal of the approach described and advocated in this book had its roots in Bell thinking. I would like to record our appreciation of the generous tradition of help to others by telecommunications people in the United States.

J. M. Harper
Amersham, June 1989

Dedication

This book is dedicated to the external cable and radio construction staff, who work in all countries and in all weathers and conditions to make it all possible.

1 Introduction

The Independent International Commission for World-Wide Telecommunications Development (the Maitland Commission) published its report in 1985. Chapter 10 of the report, embodying the Commission's conclusions and a summary of its recommendations is reproduced in Appendix 1. The core of the Commission's findings is set out in paragraph 5 of that chapter as follows:

We have identified several key elements in the joint effort for which we appeal.
— First, governments and development assistance agencies must give a higher priority than hitherto to investment in telecommunications.
— Secondly, existing networks in developing countries should be made more effective, with commercial viability the objective, and should become progressively self-reliant. The benefits of the new technologies should be exploited to the full to the extent that these are appropriate and adaptable to the countries' requirements.
— Thirdly, financing arrangements must take account of the scarcity of foreign exchange in many developing countries.
— Fourthly, the ITU should play a more effective role.

It is the aim of this book to assist Governments and administrations who have accepted the first three of these and are seeking to act upon them. Telecommunications are a recognized part of the fabric of modern society. No developing country can attain its full potential without them.

In many advanced countries telecommunications have now passed through their period of maximum rate of expansion. The administrations concerned have accumulated extensive experience of the problems that arise in this period, and of approaches and solutions to them. Britain is in this position. From 1965 to 1983 the number of lines on the British telephone network more than trebled, from 6 million to 21 million. Over a million lines were added in each of the peak years (see page xii). In the course of managing this expansion, British Telecommunications (BT) and its predecessor British Post Office Telecommunications (BPOT) necessarily faced and dealt with most of the problems of managing

1

.elecommunications development. Since 1973 BT has also organized and largely carried through a fundamental modernization of its network.

In the majority of developing countries telecommunications are less advanced. But many of them currently face, or will soon face, the problems of managing expansion and modernization on a substantial scale. The faster the rate of expansion or modernization the more difficult these problems are likely to be. This book analyses and distils British experience during and after the main period of network expansion to help policy makers and managers in developing countries as much as possible. In doing so, the book draws on BPOT experience in the 1970s and on BT experience as a nationalized industry and as a public limited company in the 1980s. The aim is to give as up-to-date and relevant an account as possible.

In most countries the telecommunications enterprise is already a substantial business. As it grows it is likely to become an important force in the economy in its own right. Historically, outside agencies like colonial adminstrations, former foreign-owned operating companies and multinational telecommunications suppliers have played a big part in shaping telecommunications in many developing countries. It is axiomatic to the approach in this book that developing countries should run their telecommunications for themselves. The modern management may enlist help or even take in partners from the advanced countries. But fundamentally it will stand on its own feet.

Chapter 9 contains a brief review of the technology, written for non-technical readers. Otherwise, the emphasis of the book is on management, commercial and policy questions rather than technical matters. Technical guidance is available from ITU Technical Cooperation Department (see Chapter 13) and from a wide range of technical publications. A list of useful publications is included in the Bibliography.

Technology in the telecommunications field and in other related fields like computing and television distribution continues to advance rapidly. The range of facilities and applications constantly widens. Later chapters draw on recent and current experience in Britain in this respect. Developing countries have real advantages over the advanced countries in this area. Because their networks and their operations are relatively undeveloped they are well placed to exploit the new possibilities without being unduly tied by the past. Chapter 16 discusses an important question to do with new services which is of special relevance to developing countries and which remains unresolved in advanced countries.

The book is divided into three Parts. Part I deals with questions of policy and management, under seven headings. These are:

— constitutional developments — changes in ownership, separation of telecommunications from Government and from Posts and admission of competition (Chapter 2);

— internal organization of telecommunications (Chapter 3);
— accounting and statistical control systems, which are a vital tool of management (Chapter 4);
— plans and planning, which are again vital for such a complex business (Chapter 5);
— charging for and pricing of telecommunications (Chapter 6);
— policy questions to do with the determination of priorities and the generation of internal capital in the circumstances of a developing country (Chapter 7);
— staff management and staffing practices (Chapter 8).

Part II deals with the technology and its applications, under seven headings:

— an outline of computing and telecommunications technology (Chapter 9);
— a description of telecommunications networks (Chapter 10);
— the role of radio, including a discussion of the application of cellular and satellite techniques (Chapter 12);
— the planning, management and procurement of network plant (Chapter 11);
— a description of the International Telecommunications Union and the Centre for Telecommunications Development and of international services and the plant they use (Chapter 13);
— a review of the telecommunications equipment used on customers' premises and a discussion of policy and management problems to which it gives rise (Chapter 14);
— a review of additional 'value-added' services provided over modern telecommunications networks and a discussion of policy relating to them in developing countries (Chapter 15).

Part III embodies a summary of conclusions, in Chapter 16.

Part I Policy and Management

2 Constitutional issues

There have been a number of important developments in various countries in recent years in the way telecommunications is constituted and organized. The purpose of this chapter is to review constitutional developments in Britain and their significance for developing countries.

Constitutional developments

The constitutional changes have affected:

— the ownership of telecommunications entities;
— the relationship of telecommunications entities to Government;
— the separation of telecommunications from posts and other activities;
— the admission of competition in the telecommunications field.

The position in Britain

In Britain telecommunications were the responsibility of a Government Department, the General Post Office (GPO) up to 1969. The GPO was also responsible for posts and for oversight of broadcasting. Telecommunications revenues were treated as part of the general revenue of the state, along with the proceeds of taxation. Telecommunications expenditures were met from moneys voted by Parliament, along with all other Government expenditures.

The telephone and the telegraph had both been introduced originally by private interests in the middle of the 19th century. They had progressively been taken over by the state. In 1912 the GPO had finally taken over responsibility for operation of telecommunications throughout Britain, with the exception only of Kingston-upon-Hull, Portsmouth, and the Channel Islands. At the same time the GPO began to publish

Commercial Accounts, which related Post Office revenue to expenditure, so that the commercial viability of operations could be determined. Separate Commercial Accounts were published for posts and telecommunications. But the Commercial Accounts were published for information only. The GPO elements in the main public revenue and expenditure account remained the formal basis of postal and telecommunications finances.

GPO finances were legally separated from those of central Government by the Post Office Act 1961. Commercial Accounts, including separate profit and loss accounts and balance sheets for the two services became the formal basis of posts and telecommunications finances.

The Post Office Act 1969 created a nationalized Corporation called The Post Office, which assumed responsibility for posts and telecommunications and which published a Report and Accounts on the usual commercial lines. On the British model, the Corporation was wholly owned by the state, but was outside Government. Its staff were not Civil Servants. The oversight of Broadcasting remained a Government responsibility.

The British Telecommunications Act 1981 enabled the Government to introduce competition in telecommunications. It also created two separate nationalized corporations — The Post Office, responsible for posts, and British Telecommunications (BT), responsible for telecommunications. BT was transferred to private sector control ('privatized') in 1984. It became a public limited company (plc). 50.2 per cent of its shares were sold on the open market; Government retained the remainder.

At the time of writing, there are two principal private sector carrier operators besides BT. The first is Mercury Telecommunications Limited, a wholly owned subsidiary of Cable and Wireless plc, which operates competing long-distance, local and private services. The other is Racal Vodaphone Limited, a wholly owned subsidiary of Racal plc, which operates a cellular radio service (see Chapter 11). BT participates in cellular radio through Cellnet, a company owned by BT and Securicor plc. For historical reasons the Corporation of Kingston-Upon-Hull, in North-East England, operates its own local telecommunications services. So far, two cable television companies have been authorized to carry telecommunications traffic on their plant.

A regulatory agency called the Office of Telecommunications (OFTEL) was established in 1983. It is headed by the Director-General of Telecommunications. His function is to oversee and regulate the activities of BT and its competitors. The effect of creating OFTEL was to separate the operating and regulatory functions which BT had inherited from its predecessor. If competition is admitted in any sector (see below) and the main network enterprise is one of the competitors, it becomes inappropriate for that enterprise to continue as the regulatory authority or to approve its competitors' customer premise equipment (see Chapter 14).

Implications for developing countries

The question of ownership and privatization is a political matter, for decision by each country for itself, and therefore outside the scope of this book. Certain countries have, however, admitted private capital into their telecommunications. In later chapters references are made to the position of such enterprises where they are appropriate.

In developing countries telecommunications has often been organized along with posts as it was organized in Britain, as part of the main Government structure. But British experience is that from all points of view — relation to central Government, internal efficiency and the interests of users — telecommunications is best treated as a business activity, financially and organizationally separate from Central Government.

There are a number of reasons for this. In the first place, the objectives and priorities of the Central Government structure are of their nature geared to the formulation and execution of public policy. Telecommunications, on the other hand, is essentially a service industry, for which the first priority is efficiency in the conduct of technical operations.

The disposition of telecommunications income is also important. As we have seen, GPO revenue was originally fed into the central finances of the state and GPO expenditures were met from the same source. Charges were treated as a way of raising revenue like taxation. The effect of such arrangements is to conceal the financial effectiveness of the operations of the administration and to deny to it and its customers the advantages of commercial operation, like ploughing back surpluses. A country which wishes to have an efficient telecommunications enterprise must be prepared to separate its finances from those of the state. A levy can still be imposed on the profits or surpluses of the enterprise if they are an indispensable source of revenue to the state, although it would be much better if the enterprise were treated for tax purposes like any other business.

In an expanding telecommunications network, capital financing arrangements are of crucial importance. Proper planning and development of telecommunications is impossible unless the enterprise has adequate and predictable capital funding for several years ahead. The central Government finances of any state are inevitably subject to complex priorities, which frequently change in the short term for political, macro-economic and other reasons. This makes it unsatisfactory for telecommunications investment to be treated as part of these finances. British experience strongly confirms that the telecommunications investment programme should stand on its own, subject to its own priorities and with scope to arrange its own funding solutions.

There are two other points concerned with financial mechanisms. The internal efficiency and viability of an enterprise can only be assessed and action taken to enhance them if they can be measured. This requires that inputs (such as, in the case of telecommunications, staff effort and investment in plant) be related to outputs (numbers of lines served, calls connected and so on). The most meaningful and convenient way to relate such inputs and outputs is in money terms. Central Government accounting structures are designed for quite different purposes, and often make it difficult if not impossible to meet the needs of an activity like telecommunications. Also, telecommunications are usually charged directly to users on a transaction basis. Knowledge of costs is essential for orderly price fixing for this purpose, even if services are to be subsidized. In a Government accounting system it would be unusual and often difficult for costs to be allocated in detail to particular services and activities in the way required for sound telecommunications costing.

The requirements of telecommunications accounting systems are discussed in more detail in Chapter 4. But as a general matter they are unlikely to be met by Central Government accounting systems, designed as they must be for quite different purposes.

Personnel management considerations are also important. Whatever happens, Government personnel management has to deal with two groups of staff with special needs and constraints. The first group are the highly specialized staff who conduct the central activities of the state and support Ministers. The second are the staff who carry out administrative processes like collection of taxes and distribution of aid.

If posts and telecommunications are organized as part of central Government their staff too must be catered for by Government in its personnel practices. In Britain the staff of the GPO, as a Government department, represented half the Civil Service. The GPO was allowed considerable autonomy to deal with its own staff matters. But in the end basic personnel issues like pay and grading required to be treated in a uniform way across the Civil Service. Yet even at that time telecommunications was a large service industry, employing substantial numbers of technical and other basic staff, supervisors and managers. Such an industry needs the freedom to tailor its pay, grading and other personnel practices to its needs. British experience confirms that this can only really be done if the industry is managed quite separately from central Government.

Even though it may be wholly state-owned, it seems clear that telecommunications should be financially and organizationally separate from the central machinery of the state. The exact basis on which this is arranged is a matter for the country concerned.

Separation of posts and telecommunications

In many countries telecommunications has been linked with posts and sometimes with other activities such as transport and tourism. While it continues to be run as part of central Government, this is a matter of the grouping of Government and Ministerial functions, outside the scope of this book. If the argument that telecommunications management should be separated from central Government is accepted, however, its grouping with posts or other activities becomes a managerial matter.

In less developed countries, especially those with large outlying areas, there have often been and may still be reasons for combining postal and telecommunications functions in the field. In sparsely populated areas communications may best be looked at as a single matter with a unified approach. For example, money can often be saved by combining the use of buildings and other facilities for the two services. But at country-wide level the clear trend in advanced countries is to separate posts and telecommunications. There are good reasons for this.

Even in countries where postal sorting operations are increasingly mechanized, as in Britain, post remains a labour-intensive business, whose activities centre on the problems of managing a large labour force and country-wide physical collection and distribution operations. Telecommunications, on the other hand, is more and more capital-intensive, and based on ever advancing and highly specialized technology.

Telecommunications is by its nature a growth business. In developing countries its problems are usually dominated by the problems of funding and of managing expansion and modernization. Not only the scale, but the range of telecommunications activities is constantly growing. The activities of posts, on the other hand, remain largely concentrated on a few traditional services. Volumes of postal traffic may expand with economic development, but it is unlikely that they will do so at a rate comparable with telecommunications expansion. Other activities like tourism and savings are even more clearly different from telecommunications.

The conclusion must be that it is preferable for a modern telecommunications enterprise to be separated managerially not just from Government but from posts, with its own senior direction up to the highest levels. This does not preclude having staff and buildings with both posts and telecommunications functions in remoter areas, where this is desirable and appropriate in local circumstances.

Even if telecommunications management is separated from postal management and from Government in this way, matters of telecommunications policy and so on will still call for the attention of Ministers and Civil Servants. The grouping of these residual Government functions in relation to telecommunications with other functions is a matter outside the scope of this book.

Competition in telecommunications

Like privatization, the question of whether competition should be admitted into telecommunications is a political matter for decision by

Governments. British experience, however, suggests certain conclusions about the effects of competition which it may be helpful to record.

Potentially, competition has an important role in the development and exploitation of telecommunications technology. Under monopoly conditions the whole community is dependent on the operator for the selection and promotion of new developments. Telecommunications techniques are advancing very rapidly. New possibilities for apparatus and services are constantly emerging. In such a situation in an advanced society it is impossible to be sure that a monopoly operator, however vigorous and alert, will identify or will have the resources to exploit all the new ideas of potential value that may be appearing.

Developing countries are, however, in a better position. They can take advantage of the exploitation of the technology in advanced countries without necessarily having to undertake technical development or the devising of new applications for themselves. All that is required in their case is vigilance and readiness to seize on new developments appearing on the world scene when they are relevant.

From the point of view of Government and the public it is much more difficult to be sure that monopoly activities are being efficiently run than competitive activities. Detailed Government scrutiny of operating results and performance, including international comparison with other operating entities, and surveys of consumer opinion can provide a degree of check on monopoly operation. But none of them is as effective as a spur to efficiency as competition.

On the other hand, there are real limitations on the scope for competition, especially in the circumstances of a developing country. In the first place, even in Britain and other advanced countries it has only proved possible so far to introduce significant competition in customer apparatus, network services (VANS — see Chapter 15), long-distance telecommunications and cellular radio.

Local service is at the core of telecommunications management problems. Yet in Britain, as in other advanced countries, competition in local telecommunications is as yet confined to very limited, high-density geographic areas, and seems unlikely to become anywhere near universal. The principal reason is the very high cost that would be involved in duplicating local cable distribution networks; and the doubtful return that the investment would earn, even if revenue were supplemented by distributing television programmes (see Chapter 12).

So far as long-distance communications are concerned, there are important considerations concerning economies of scale. At the scale of operation in Britain or the United States competition in long-distance inland telecommunications may be expected to be economic in suitable situations. But so far as can be judged from British experience, at the present scale of operations of most developing countries it must be doubtful whether any kind of long-distance network competition involving the duplication of network facilities could be economic. Certainly a proposal to introduce it would call for thorough study of the economics and of such matters as the effect on the environment of duplicating radio masts before a decision was taken.

So far as can be judged from what has happened in Britain, therefore, in the circumstances of most developing countries the only areas where there is likely to be serious scope for competition are provision and installation of customer premise equipment (CPE) and network services (VANS). These are discussed in Chapters 14 and 15. Network operations should probably remain a monopoly. The operation and improvement of the network would then be the priority task of the national telecommunications enterprise.

Attention should be drawn to one important consequence of this conclusion. The effect of competition is to provide a spur to the efficiency with which the activities it affects are performed. If there is no such spur the stimulus to efficency must be found in other ways.

The BT experience is that in a country-wide monopoly or near-monopoly telecommunications enterprise this stimulus must be provided by the top direction of the enterprise; and that the management and control systems must transmit it effectively out through the field structure to individual supervisors and their staff. These are not easy things to do. They are made more difficult because of the parallel need for devolution of decision-taking. The most important purpose of the organizational structures and control systems described in the rest of this Part is to give effect to and reconcile these two equally important objectives.

Conclusions

To summarize:

— There has been a steady process of constitutional and organizational change in telecommunications in advanced countries in recent years.
— The question of state or private sector ownership of telecommunications is a political matter outside the scope of the book.
— British experience suggests that telecommunications are best organized separately from central Government, even though they may be state owned; and that telecommunications should be separated managerially from posts.
— The introduction of competition in telecommunications has the potential to provide a spur to efficiency and exploitation of the technology. There are, however, important practical difficulties about introducing it in developing countries. There may be scope for competition in customer apparatus (Chapter 14) and VANS (Chapter 15), leaving the network as the priority for the national enterprise.

— In a monopoly operation the top direction of the enterprise must provide a major internal spur to efficency of operation. The rest of this Part is concerned with mechanisms which have this as a major objective.

3 Organization

As a modern business enterprise with monopoly or near-monopoly responsibilities, telecommunications requires a properly structured internal organization. The provision of telecommunications services involves many activities, such as engineering, operating and maintaining plant, finance and personnel management, which are common to many other kinds of undertaking. The general principles of modern business organization are dealt with in a number of excellent publications (see particularly Peters and Waterman, 1982). But telecommunications has particular characteristics which impose special demands on organization.

Many permutations of telecommunications organization exist round the world. It would be impossible to discuss or draw conclusions about the full range. But BT experience has suggested that certain issues are of particular importance. These are discussed in this chapter.

Overall structures

In most undertakings, however large, capital plant and engineering operations are likely to be concentrated at a relatively small number of sites. In telecommunications, however, they are country-wide. Telecommunications facilities require to be provided in all populated areas. Some of the most difficult problems arise when extending service to sparsely populated areas which often have awkward terrain. Telecommunications organization must cater for these requirements.

There are other considerations. The telecommunications network is essentially an enormous machine for the making of any of an astronomical — and steadily growing — number of temporary individual connections between consumers throughout the world, and between them and a constantly growing variety of services and facilities.

In any organization the number of things likely to go wrong rises as at least the square of the number of individual transactions the organization has to handle. Telecommunications handles vastly greater numbers of such transactions than any other public utility. Each call is really a transaction in its own right. Also, telecommunications service is an immediate and personal thing. The result is that through no fault of its own a telecommunications enterprise will always have many more potential customer service problems than a public utility like, for instance, electricity. To satisfy the public, it needs to catch most of them before or when they arise. It follows that telecommunications has a special need for a structure which will guarantee that local operations are closely responsive to customer needs and opinion.

The general questions that arise in meeting these requirements can most easily be discussed by referring to three skeleton organizational diagrams, as illustrated in Figures 3.1, 3.2 and 3.3.

The terminology used in these diagrams is chosen to be self-explanatory. There is one exception. Historically the BT structure included a 'traffic' function organized in a separate Civil Service hierarchy. Traffic staff were concerned with the forecasting and routing of calls, with the management of the operator force and with all matters to do with call service to customers including such things as dialling code lists. In BT most of these functions have now been absorbed by other groups. A separate traffic staff no longer exists. For example, management of the operator force is a distinct activity in its own right. But a call service function is required in any enterprise to handle customers' instructions, directories, and all matters to do with circulation of calls and call service complaints. The diagrams provide for this activity as 'Call Service'.

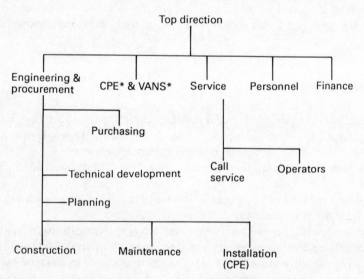

*Customer Premise Equipment and Value Added Network Services (see Chapters 14 and 15)

Figure 3.1 Typical skeleton structure A (fully centralized structure) (line functions are shown in capitals)

Structure A

The distinctive characteristic of this structure is that there is no coordinating general management except at the very top level. Each specialism is distinct in both the field and headquarters organizations up to the level immediately below the highest.

Prior to the 1940s, GPO Telecommunications were organized broadly on these lines. GPO experience was that the arrangement had serious disadvantages, especially in relation to field operations. All important decisions and all decisions involving more than one specialism had to be referred to high levels in the central headquarters. This frustrated local supervisors and managers, destroyed their initiative and slowed the whole pace of operations.

Another difficulty with this structure is that there is no organizational mechanism for singling out for attention operations in particular geographic areas. This is an important defect. The needs and problems of large towns and rural provinces are usually distinctive, and deserve dedicated management attention. Also no one person is available to represent all aspects of the enterprise in local and provincial communities.

+ Typically Exchange and transmission systems
* Typically Local Distribution Network only

Figure 3.2 Typical skeleton structure B (partly decentralized) (structure line functions are shown in capitals)

Structure B

The distinctive difference from Structure A is that field operations are grouped under the control of field general managers who take responsibility for the whole provision of service to the public in their areas. As in Structure A there are still a series of specialist functional Departments at Headquarters. These Departments are responsible for a number of executive matters like network planning to the extent that this is centralized — for example, for the long-distance network, procurement, introduction of new customer services and apparatus and so on; for the preparation of detailed instructions for use by the field units; and for specialized oversight of and advice to those units, for example, on technical and commercial matters. The top direction may have a small operations and planning staff unit to assist it with line oversight of field general managers and central planning.

The field general managers have executive groups responsible for plant construction, plant maintenance and so on as indicated. These groups are under the line control of the general managers, but look to the Headquarter Departments for oversight and support in their specialisms.

The advantages of this structure are that matters wholly within the territory and authority of a field unit can be decided on the ground, without reference to Headquarters. The field general managers are formally answerable for operations and performance in their commands, and represent all aspects of the telecommunications enterprise in their local or provincial communities. Specialist staff concerned with more complex technology and similar matters remain concentrated at the centre. This is an important advantage in developing countries, who may be short of specialists.

One difficulty with Structure B concerns the relationship between the specialist Departments at Headquarters, the related functional groups in field units and the field general managers. The field functional groups may experience divided loyalties between their local general managers and their 'parent' Headquarter Departments. The fact that these groups are required still to follow rule books laid down by central specialists can inhibit local initiative and decision-taking. The combined effect can be to complicate and delay decision-taking and to blur lines of responsibility.

The BT structure was of this type from the Second World War through to 1983, including the main period of expansion of the network, except that it included a Regional Headquarters tier (see below). By the end of the period, however, it was clear that the growth in field operations and the special needs of telecommunications for a responsive field structure made further change desirable.

Structure C

In this structure field operations are grouped as in Structure B, but there is the greatest possible devolution of functions and responsibility on to the field general managers. The only matters retained at the central

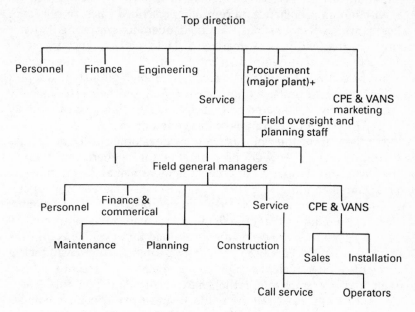

Figure 3.3 Typical skeleton structure C (fully decentralized structure) (line functions are shown in capitals)

headquarters are those which of their nature must be conducted centrally. These include, for example, overall financial planning, network layout, major plant procurement and national strategies for the introduction and marketing of new products and services. It may also be preferred to retain central control of particularly sensitive matters which could in theory be handled locally, like wage and price fixing. Otherwise the field units are left to manage their own affairs, subject to answerability for financial and other results each year through mechanisms described in Chapters 4 and 5. Detailed technical manuals may still have to be prepared centrally, but otherwise national 'rule books' are cut to a minimum.

In the most business-oriented application of the structure, the field units are treated as the equivalent of wholly owned subsidiary companies, managed and overseen primarily in terms of profit and loss accounts and balance sheets. If responsibility is to be convincingly devolved they should even control their own wages and prices. In a country which is prepared to privatize public utilities, such subsidiaries could of course become independent private sector companies in their own right, and thus very firmly rooted in the towns and provinces they serve.

This structure draws on recognized modern management practice. It has important advantages. Most decisions can be taken close to the workforce, with minimum reliance on centrally prescribed rules, or reference to Headquarters. Responsibility for field operations is vested unambiguously in the field general managers. Field staff are encouraged to identify wholeheartedly with their local units. The role of the departments in the central Headquarters is equally unambiguously to support and advise the field units, without encroaching on their responsibility and answerability. Compared with B overheads should be markedly reduced. BT's present structure is of this general type.

Nevertheless, from the point of view of developing countries C has certain disadvantages. To work properly it requires general managers to lead the field units who are not only competent managers but have an adequate understanding of all the specialisms — engineering, operations, finance, personnel and so on — and a grasp of modern business practice. It also requires that each field general manager is supported by senior managers and professional staff trained and experienced in the various specialisms. Finally, it requires people in the Headquarter Departments with sufficient insight to understand and accept that they are in an advisory 'staff' role, rather than a controlling 'line' role. These are not necessarily easy requirements to meet even in an advanced country. They may be difficult, if not impossible, to meet in developing countries, who simply may not have enough trained and experienced people.

It must be emphasized that this outline is concerned only with the principles of three possible approaches to a problem with many solutions. A choice of solution can only be made and the detailed structure appropriate in a particular country determined in the light of local conditions. Generally speaking, structures based on A are likely to be suitable only for small enterprises, with limited geographic territory, or at an early stage of development. Structures based on C, on the other hand, are probably best suited to large countries, with well-developed networks, well-defined provincial centres and a sufficient supply of trained and competent general managers and specialists. B is perhaps the best model for enterprises involved in or preparing for major expansion in most countries.

I served as a field line manager (Regional Director) in the BPOT structure based on B from 1969 to 1972 and as a Headquarters line manager from 1972 to 1983. In my experience the structure worked well, with three reservations. First, there were tensions between Headquarter Departments and field unit specialists referred to earlier. These were never wholly satisfactorily resolved, though they were usually contained. Second, and more important, the central top management was never able to motivate and get feedback from the field force properly. Third, we were never able to eradicate a disposition in some field units to look more than was needed to Headquarters for leadership and decision-taking.

Nevertheless, my experience suggests that such a structure could work better than any other in the circumstances of a developing country with an expanding network.

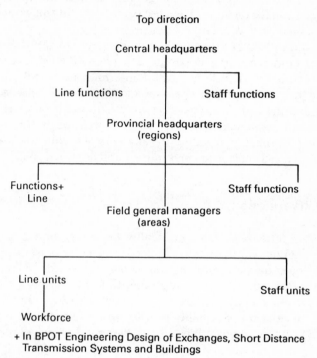

Figure 3.4 Three-tier organization

$\bullet$ Series - rank $\bullet$

Tiers of organization

As described above in Figures 3.1 – 3.3, Structures B and C both assume that the requirements of the territory concerned can be met by a single tier of field organizations. In certain advanced countries, including Britain until recently, there were two tiers of field organization (Figure 3.4). In Britain the middle tier comprised intermediate headquarters (called Regional Headquarters) based in large provincial centres and led by senior line managers (Regional Directors). The third tier comprised Areas, led by their own General Managers. The principal functions of the Regional Headquarters were to reduce the span of control of national headquarters; and to provide concentrations of network and other technical and specialist expertise closer to operations on the ground than the national headquarters. In a country of the size and density of population of Britain and in conditions of rapid analogue network expansion there was a recognizable role for units of this kind.

BT has recently restructured its field organization to eliminate the regional tier, and to reduce the number of field units in the lower tier from fifty to thirty. The new field units are called Districts and are grouped in three Territories. Each Territory has a Director, who is part of the Central Headquarters, responsible for oversight of and communication of policy to the Districts in his territory. He is not part of field line management and has only a small support staff.

The question whether a two- or three-tier field organization is appropriate in developing countries depends very much on local circumstances. There are important objections to a full intermediate tier. It must add to costs and to the length of the management chain between central direction and the units where actual work is done. It involves an extra stratum of specialist staff who may be in short supply. If it is justified anywhere it is probably in geographically extensive countries, with well-defined provincial centres of Government and administration.

The quality function

BT recent experience is that the quality of service and workmanship deserves as much emphasis in the telecommunications service industry as it has come to receive in the manufacturing sector. BT now has a specialist quality staff Director reporting to the Chairman at corporate level. His responsibility is to oversee the creation of quality organs and the observance of proper quality practices throughout the Corporation. Modern BT quality philosophies are discussed in Lomas (19::).

Network organization

The discussion of possible field structures in the preceding paragraphs did not deal in detail with organization for the planning and management of the network of lines and exchanges. Some complex questions arise in this area.

The organizational requirements are different according to whether the network is primarily analogue or digital (see Chapter 10). In an analogue network each exchange and inter-exchange link is planned and managed separately from the technical point of view. The work of planning and operating the network can be allocated among the various formations according to the wider logic of the organization. In BPOT in the 1970s, for example, the long-distance transmission network and the associated exchanges were planned in the Central Headquarters; the Provincial Headquarters (Regions) planned the short-distance inter-exchange (junction) network and local exchanges; and the local units (Areas) planned the customer distribution cable network. Major exchange plant and long-distance transmission plant was supplied and installed by contractors.

The Areas were responsible for direct (BT own staff) labour construction and maintenance work on all network plant.

In a predominantly digital network the technical situation is different. In a sense the digital network, including both local and long-distance exchanges and all inter-exchange circuits, is a single entity from the technical point of view. Its functioning depends on instantaneous network-wide interaction between control computers, which are likely to use advanced computing technology like artificial intelligence before many years are past and other high-technology plant. The functioning of the various digital switching and transmission entities throughout the network requires to be synchronized in a very precise way. The flow of traffic and measures to contain congestion and breakdowns are actively managed from a central point. Also, the number of staff working on the floor of exchanges and transmission stations or their equivalents is much fewer than it used to be. From the engineering point of view, therefore, it is arguably both desirable and practical for all planning, construction and maintenance work on exchanges and the inter-exchange network to be under unified control.

The local distribution network will incorporate the equivalent of switching functions involving high-technology plant. It too is likely to need to be engineered in such a way that it is capable of extensive interaction with the central network.

On the other hand, SPC digital networks are inherently capable of organizational subdivision at defined 'interfaces'. For example, to discharge their role, digital control processors (see Chapter 9) must present uniform interfaces to one another whoever runs them. In present networks they exchange routing and control information by using a uniform signalling system whose technical characteristics have been agreed at world level known as CCITT No 7 (see Chapter 13). Two processors can therefore interwork fully even though they may belong to totally separate enterprises.

Thinking in advanced countries on this question is still evolving. In some of these countries, depending on the circumstances and constitutional approach, it may prove desirable to distinguish organizationally between the long distance network, the local exchange and junction network, the local distribution network and/or the actual provision of services to customers. Different numbers of operators and competitive or other regimes may be desired at these various levels.

At the time of writing such questions are being canvassed in Britain. For example, there may prove to be a case on constitutional and organizational grounds for separating agencies concerned with long-distance communications from agencies concerned with local switching and distribution; or for grouping local and long-distance switching and transmission together but separating out local distribution and separating that in turn from the provision of services to customers.

The question whether such approaches are appropriate in developing countries will depend very much on circumstances and the desired constitutional approach. In a large or particularly advanced country it may be appropriate to consider carefully whether such changes would be desirable in the immediate future. But in the majority of countries, with less evolved systems, the traditional unified approach will probably continue to be right, at least for the present.

'Long Lines' organization

Structural distinctions of this kind have been made in the United States for many years. Within the Bell System up to 1984, long-distance and international service was the responsibility of AT&T Long Lines, and local service was the responsibility of twenty-two local operating companies. Long Lines and the operating companies were separate subsidiaries of the parent AT&T Company. Since 1984 the local companies have been totally separated from AT&T, and are now grouped in seven Regional Bell Operating Companies (RBOCs). Long Lines remain a function of AT&T.

The question of whether to introduce a 'long lines' organization in BPOT was under review for a number of years. Certain steps to separate long-distance and local service were taken following the 1969 reorganization. Engineering planning and provision of long-distance transmission links and exchanges became the responsibility of a separate Network Planning Department at headquarters. Day-to-day operation and maintenance of the network as a whole, however, remained the responsibility of the combined field organization with no distinction between long-distance and local plant.

In 1982, in face of the prospect of long-distance carrier competition (Mercury), BT created a National Networks Division. Within this division Trunk Services had responsibility for the long-distance transmission network and for all exchanges exclusively concerned with long-distance traffic, including maintenance and exploitation and such matters as tariffs. Trunk Services had its own field staff and structure, quite separate from those of local operations. National Networks also incorporated Specialized Services, responsible for major business facilities like long-distance leased lines (private circuits), telex and packet-switching and dealing direct with major customers.

BT modified this structure again in 1986. Trunk Services' field staff and structure remains separate from local operations, but at top level trunk services are grouped with local services in a Division called United Kingdom Communications (UKC). Strategy for the exchange and transmission network as a whole is now the responsibility of a single central unit.

Few developing countries are comparable in size with the United States, and none has the special regulatory and legal background. The

United States parallel is probably therefore of limited relevance in their case, except perhaps in countries with exceptionally large territories.

Organization within field units

There are many different ways of organizing field units internally. In organizations based on Structure A the field organization would be functional. For example, in a given locality there would be an engineering group, a service group, a sales group and a commercial and administrative group. The engineering group could be further sub-divided into planning, construction and maintenance groups, and so on. Each group would report to its parent Department at the Central Headquarters.

In organizations on the lines of B or C there is a choice between this kind of functional organization and a 'product' oriented organization. For example, the field work of constructing, maintaining and operating network plant and connecting customers' premises might be concentrated on a single group with its own technical, commercial and administrative staff. A similar group might be responsible for selling and installing customer apparatus.

A functional organization has certain advantages, especially where staff with technical and other specialist skills are in short supply. In such circumstances it may be essential to concentrate engineering or computing expertise in a single group rather than to disperse it over a series of 'product' groups. It is also much easier to arrange support from specialist groups in higher formations in a functional structure. For example, a customer cable planning specialist in National Headquarters will find it much easier to interface with a corresponding specialist in the field than with someone with multi-disciplinary responsibilities.

But generally speaking 'product' organizations are probably to be preferred under modern conditions. This is particularly true in fields where competition has been introduced. If competition is admitted in CPE, for example, a dedicated CPE group can be created, with all its own technical, commercial and support staff. A compact group of this kind will be much better placed to take on its competitors, who are likely to be organized on similar lines, than a large field unit organized on functional lines. More generally, staff are likely to identify most closely with their work if they feel that the immediate unit within which they work is master of its own destiny and will get clear credit for its acheivements. In a functional organization responsibility and credit are diffused.

One important feature of field unit structure should be noted. BPOT and BT experience has underlined the strong desire of customers to have a single point of contact with a telecommunications enterprise for all their day-to-day business. They object to a situation in which they have to know before they start to whom to refer a query about, say, the bad service they are getting and to distinguish that from a requirement for a new line or facility. They also object strongly to being referred from the engineering to the commercial group because the organization is not geared to their particular situation. A customer who wants an additional line, for example, may also want an answering machine, and may at the same time have service problems. He would very much prefer to be able to make all three requests to a single person whom he already knows to having to approach three separate and probably strange groups.

In a business as complex and specialized as telecommunications it is not easy to meet this requirement. If a proper maintenance service is to be given, for example, it may be essential to have several engineering sub-groups, one of which specializes in exchange maintenance, another in customer equipment, a third in customer cables, and so on. Equally it may be undesirable that such technical staff should have to use time on dealing with callers, record-keeping, the making out of accounting vouchers, and so on.

Some of these problems can be resolved by setting up a properly designed computer system which can hold all a customer's records in a way which can be accessed by any member of staff. Enquiries from customers on any subject can then be pursued by whichever group of staff is approached first.

Whatever is done about use of computers or internal organization the best solution will always be to identify a single person to handle all a customer's business and to tell the customer his name and telephone number.

Buildings and accommodation

An enterprise which is expanding its network is likely to require a substantial building programme (see Chapter 12). The GPO originally had its own staff of architects, but from the 1950s to the 1970s all substantial building design, construction and maintenance work was carried out by the central Government buildings agency, the Property Services Agency. Since then BT has progressively taken over respon-sibility for maintenance of buildings itself and it now commissions architects and contractors directly. It has its own estate management staff.

The practice in developing countries will depend on local circum-stances. Where a central Government buildings agency exists there will often be strong arguments for having it handle buildings work for the telecommunications enterprise. But telecommunications buildings usu-ally have special design requirements and need special attention to ensure they are completed on time (see Chapter 12). BT experience is that it is more satisfactory for the enterprise to handle its dealings with architects and contractors itself, if this is reasonably economic.

Building work is probably best organized within the enterprise as part of the engineering function. It will usually be appropriate to handle the

main building programme at Central Headquarters or Provincial Headquarters if these exist.

Directories and directory enquiry

Up-to-date directories are very important for the efficient operation of the
service. If the directory is inadequate customers will be forced to consult
directory enquiry or normal assistance operators. This must cost money if
an adequate operator service is to be given. More generally, poor number
information services will result in an increase in the number of ineffective
calls on the network, with a consequent loss of revenue and increase in
costs. Directory enquiry work is an increasing element of operator service
costs in most enterprises. Most advanced countries including Britain have
now introduced computers for the use of directory enquiry operators.
The technical problems in this area took many years to solve, but
satisfactory directory enquiry computer systems are now on the market.

The collection of information for subscribers' entries in directories can
only be carried out within the enterprise. It is normally arranged to occur
automatically as part of the process of provision of service. In a computer-
based organization the text of the directory entry can be an automatic by-
product of the provision procedure.

The compilation, printing and publication of directories may be carried
out by the enterprise. But it will often be convenient to arrange for
compilation and printing to be carried out by an outside organization —
either the state printing and publishing organization or a private firm.
The precise arrangements will depend on practices in the country
concerned. Directories are often distributed by the postal enterprise,
although the telecommunications enterprise can of course do this itself or
engage outside contractors.

In Britain and France steps have been taken to give customers direct
access to the computers maintained for the directory enquiry service.

New products and services

New terminal apparatus products and new network services play an
important role in modern telecommunications. They pose organizational
problems. Decisions about introducing them and about strategies and
tactics of promotion fall naturally to central headquarters. At first sight
the business of selling them, installing and maintaining the apparatus
and making the services available to customers falls equally naturally to
the normal field organization. But this work often requires specialized
knowledge which may not be available in the field. Conflicts of priority
may also arise if the work has to be carried out by staff also responsible for
pressing day-to-day work connected with basic telephone service.

As an alternative the work may be carried out by a special, centrally organized sales and engineering force. The advantage of this arrangement is that it concentrates specialist knowledge which may be at a premium, and that such a force can be exclusively motivated to secure and carry out the new kinds of business. On the other hand, the interest of the new products and the prestige of being concerned with them can act as valuable motivators for the main field force if it does the work. Also, the case is made in Chapter 15 for a policy of spreading the benefits of new network services as widely as possible. Such a policy will obviously be helped if staff who can offer and provide the new services and facilities are located throughout the country and parented on the normal field organization. These questions can only be decided in the light of local circumstances.

The question of functional or 'product' organization discussed above in the context of field units also arises in connection with the way work on new products is organized in the Central Headquarters. The choice is between a number of small groups, say, for telephones, PABXs, ancillaries, VANS and so on, each with its own marketing, engineering and support staff and a functional organization dealing, say, with all CPE. There are strong arguments in favour of the former, but it has its own problems. Technical or market developments may mean that a small group gets out of its depth. It is also likely to be very difficult for a small group, with its attention concentrated on day-to-day sales, to spare time for strategic thinking about new products. On the other hand, a functional group is liable to be slow-moving and bureaucratic in its approach and divorced from the situation on the ground.

The best solution is probably to organize tactical sales and marketing, technical support for customers and so on in small groups, but also to have a strategic unit responsible for market and product planning over broader sectors closely associated with but separate from the tactical groups. Such an approach will only be possible, however, in larger enterprises. Whatever approach is adopted, in this particular area top management needs to watch closely how matters develop and be ready to change the organization fairly frequently.

International telecommunications

The discussion so far has been concerned with the organization of inland activities. International services are important in developing countries and their organization deserves close attention.

Two kinds of international activity are involved. The international service will normally have its own plant — exchanges, submarine cable terminals and earth stations and its own operators. The staff concerned have to be managed and the plant has to be planned and installed. The enterprise will also have to negotiate with foreign administrations and with the ITU* about commercial and technical matters. In BT both functions have been carried out by a single international entity since the 1960s. All experience indicates that this is the best arrangement. British Telecom International (BTI) is now a separate Division of BT with its own Board Member and Managing Director.

Organization within the international division is a specialized subject which it is not possible to pursue here in detail. Certain points should, however, be noted. International negotiation activities are likely to require the head of the international division to be abroad for substantial periods. It is important that he has a deputy or equivalent who can oversee the day-to-day provision of international services and deal with other matters in his absence. In BPOT, when the structure based on Structure B was in force, a distinct operational unit was created within the international division. This unit ranked as a Region within the main structure of BPOT, even though its head reported to the head of the international division.

Conclusions

Formal organization is important in a telecommunications enterprise. The way work is grouped can have a formative effect on the service given to customers, on the successful application of the technology, and on costs. Telecommunications is a high-technology business and it is important that senior management includes a sufficient number of people with professional engineering backgrounds. Such matters deserve close attention by top management. In some cases consultants may be valuable.

Informal organization and habits of thought and such matters as the physical disposition of staff within and between buildings are as important as formal organization. It is important that everyone remembers as he does his job that telecommunications has to be treated as a single organism. Each part interacts with all the others. For example, engineering arrangements for call charging on one exchange must interact properly not only with all the other exchanges (in detecting when calls are answered, for example) but also with the accounting and billing systems. The various parts cannot be optimized in isolation.

In recent years many advanced country administrations have found that they have to modify their inland telecommunications structures at frequent intervals to meet the changing demands of the technology and the market. Developing countries must expect to have to adapt their

---

* The ITU is described in Chapter 13. Some ITU activities such as tones heard by customers, numbering schemes and so on also affect inland services.

organizations as they grow and as they introduce new services and other telecommunications applications. But major reorganizations are bound to cause dislocation, cost money and consume resources. It is therefore important that whatever structure is adopted is as well adapted as possible to expand and take on new activities, without needing to be radically altered too often.

Summary

To sum up the chapter:

— The basic questions that arise in organizing a telecommunications enterprise have much in common with those of major businesses generally.
— There are many possible permutations of telecommunications organization. It is not possible to discuss all of them. But British experience suggests that there are certain organizational issues of special importance for telecommunications.
— Britain has experience of three basic approaches to the problem of the relationship between the central headquarters and field operations, of which Structure B is probably best suited as a basis for organization in the majority of developing countries.
— A two-tier organization of Headquarters and field units is probably right for most developing countries. A third tier, involving groupings of a number of field units at provincial or conurbation level may be justified in some circumstances.
— In countries where the network is on the way to becoming fully digital it may be right to review whether any or all of long-distance network operations, short-distance network operations, the local distribution function and provision of actual services to customers should be separated out from one another.
— In the circumstances of most developing countries it is, however, unlikely to be appropriate to create a separate inland long-distance communications agency on the lines of AT&T Long Lines in the United States.
— Within field units, 'product' group organizations are probably to be preferred in a modern telecommunications enterprise.
— Customers attach importance to having a single point of contact with a telecommunications enterprise to which they can refer all their business. The problems this requirement poses can be resolved by introducing a suitable comprehensive computer system for all customer records, and notifying customers of a single person within the organization whom they can approach on any subject.
— The arrangements for selling and for engineering work on new apparatus and services give rise to organizational problems which will have to be addressed in the light of local circumstances.

— The organization of international services poses its own problems, which deserve careful attention in the circumstances of developing countries.

4 Accounting and control systems

To be properly run an enterprise must have accounting and statistical control systems which match its requirements and characteristics. To take a simple example, in telecommunications it will often be desirable as part of the process of price determination to know whether a particular sector — say, rentals for telephone instruments — is making or losing money. There is no way this can be done unless the accounting system has been so designed that instrument rental revenue and costs can be isolated from the rest of the finances. A management fixing prices without such information is doing so in the dark.

BPOT, like telecommunications enterprises in many countries, inherited an accounting system designed to meet the requirements of central Government. This system could not provide reliable information at the level of detail in the example in the preceding paragraph. British experience is that such a system is likely to require radical change if it is to be made to match the requirements of telecommunications.

The purpose of the first part of this chapter is to inform non-specialist administrators and senior managers about requirements for an accounting system for a telecommunications enterprise organized on the lines advocated in Chapters 2 and 3. It is not intended as a detailed treatment for specialists. Non-financial statistical controls are dealt with in the later part of the chapter.

Accounting and statistical systems are formative for management. They deserve close attention up to top levels in the enterprise.

Telecommunications accounting systems

The accounting system of a telecommunications enterprise has to fulfil several different requirements. First, it must enable the financial performance of the undertaking as a whole to be assessed — for example, whether it is making or losing money, whether costs are being controlled, whether prices need to be raised and if so which and when, and so on.

Government, shareholders, if there is private capital involved, and the top management of the enterprise itself will all want to be able to do this.

Second, it must guarantee that cash, stocks and transactions of many different kinds are accurately recorded and brought to account in a manner calculated firmly to discourage dishonesty.

Third, it must enable top management to make financial plans and budgets in advance, to monitor developments against these plans and budgets as they take place and to take corrective action; and to make all necessary price adjustments, including adjustments between sectors like calls and rentals, on a firm factual basis.

Fourth, it must provide the basis for review after the event. In a state-owned enterprise, for example, Government is likely to want to call the top direction of the enterprise to account at each year-end in respect of its performance during the year.

Fifth, if the organizational approach of the enterprise requires the financial performance of subordinate units to be assessed and answered for by their managers, the accounting system must enable this to be done. If operations in a particular tract of territory are to be treated as a profit centre, as they might be in Structure B or C in Chapter 3, the accounting system must permit income and expenditure to be isolated and accurate profit and loss statements and balance sheets to be prepared for that territory.

Sixth, the system must make it possible to monitor cash flow — for example, to make sure that there is enough cash available to pay wages and bills, and that excessive uninvested balances, with consequent loss of interest, are not allowed to accumulate.

In general, modern accounting systems of the kind used in major firms in most countries are well adapted to meet these requirements. The first requirement for an enterprise which does not have such a system will be to recruit professionally qualified accountants of the calibre required to introduce and operate it. This is not necessarily an easy thing to do and it may call for careful attention at the highest level.

Technical accounting problems like the correct treatment of inflation arise and have to be faced in all businesses. They pose few problems peculiar to telecommunications. But there are likely to be practical problems in introducing a business accounting system in many telecommunications enterprises. These will include the elimination of practices based on earlier arrangements — for example, public finance accounting practices where telecommunications was previously run by the state — and the conversion of records and data from the formats of the old system to those of the new.

It is extremely desirable that when a new system is introduced it is built round the fullest exploitation of computing and telecommunications technique. Thus not only should data be held on computing databases structured in a properly thought out way; but new data about man-hour consumption, cash purchases and so on in field units should be input directly to these databases from terminals, including mobile radio terminals in the work-place.

Any accounting system is only as good as the records and vouchers for individual transactions on which it is based. If the details of customer apparatus for which rental is to be charged or if the record of engineering manhours expended in maintaining a particular exchange are carelessly or misleadingly recorded, for example, no amount of subsequent processing can put them right. If they are wrong, the top-level accounts and statements based on them will also be wrong. In a business like telecommunications, involving very large numbers indeed of small transactions often carried out by unsupervised or partially supervised staff at locations spread throughout the country, the chances of error or falsification are very high. Special attention is needed to make sure that they are minimized.

Telecommunications plant also presents special problems. A great deal of telecommunications capital investment involves a large number of relatively small tranches of plant, like cables and radio stations. Even exchanges and buildings frequently comprise several generations of investment and plant, each requiring distinctive treatment for depreciation and capital charge purposes. If a proper accounting system is to be operated, each unit of investment must be recorded and the record must be held in such a way that the unique history of the asset concerned is available. In practice, this can only be done by a computer-based system of fixed asset registers. In such a system each tranche of plant, however small, is uniquely identified, and the date and value at which it was capitalized are recorded.

One more technical point should be noted. Good modern accounting systems are almost invariably based on double-entry book-keeping techniques. In telecommunications a decision has to be made on how close to the workplace full double-entry records are maintained. BT experience indicates that this should be done as close to the work-place as is economic — say, at field general manager unit level in Structure B.

It will often be helpful to employ suitably qualified consultants in the design and adaptation of telecommunications accounting systems. It is important that they take account of the requirements set out above.

Statistical returns and controls

Financial controls of the kind described in the preceding section are an essential tool of management, but they have limitations. A financial measure of the value of orders for service received and processed, for example, says nothing about the number of lines connected to the system as a result, what the backlog of orders is, or how long they took to process. Yet such information is essential, particularly in conditions of high growth and potential shortage of capacity. Information on call revenue says nothing about the number of local and long-distance calls or their average length. But such information is needed for many purposes including planning the expansion of the network, and so on.

There are other important limitations. Financial information cannot take direct account of the quality of service given. Yet from the point of view of management quality is just as important as cost. And in most telecommunications administrations information about and appreciation of the significance of results in financial terms is likely to be confined to senior management and financial specialists. Controls expressed in terms directly familiar to junior managers and staff, like number of orders delayed or proportion of calls failing, are essential if they are to do their jobs properly and be effectively motivated.

On the other hand, it is important that non-financial statistics are kept as simple and few in number as possible. Too many statistical controls will produce resentment and confusion among subordinate managers and staff, and will hinder rather than help senior management.

Basic statistics

The statistical requirements of each enterprise will vary with its circumstances. The statistics needed will generally be of two kinds — global statistics, like the number of working lines on a network, and control statistics, such as the percentage of local calls failing due to the enterprise.

Experience suggests that it will be essential to publish and circulate a basic set of statistical information within the enterprise at regular intervals, and that this should include the following:

Size of network and number of calls

Number of working telephone and telex main lines.
Number of working telephone stations.
Number and average duration of each of local, long-distance and international telephone calls; and of domestic and international telex calls.
Number of public call offices.

Demand for and supply of service
Number of applications for main lines received. Number of applications outstanding over specified period, say, two months (the waiting list).

Quality of service
Percentage of local, long-distance and international calls failing due to the enterprise.
Average time to connect calls to the Assistance Operator.
Average time to connect calls to the Directory Enquiry Operator.

Number of staff
Total number of staff.
Number each of office, engineering, operating (telephonist) and management staff.

In some cases such as the number of outstanding applications, it will be desirable to have monthly figures. In others, for example, call failure statistics, only annual figures will be of real significance. In BPOT financial information and operational statistics were published in a combined Monthly Report.

Other statistics may need to be circulated internally in the same way, according to the circumstances and management problems of the enterprise. For example, where there is a long delay in providing service it may be desirable to monitor the average time taken to deal with applications; where it is desired to control labour costs, it may be desirable to measure manpower productivity (for example, total man-hours per station — see Chapter 8); and so on. A full list of the variables monitored centrally in BT in 1987 is given in Appendix 2.

In designing a statistical system it is important that the required data can be readily and reliably produced. It is no good calling for figures that cannot be made available. In many cases computing techniques can be exploited. For example, if the process of order-taking and processing is computerized, a wide variety of statistics on order intake should be available with little extra effort. If stored programme-controlled exchanges (Chapter 9) are in operation, they should be able to provide a similar variety of statistics on the number of calls made, with extensive detail on the distance and duration of calls if required, and the number of failures. Specialized computers are available which can provide a statistically accurate measure of the quality of call service by making test calls to a predetermined pattern throughout the twenty-four hours. Where computers are not available statistics will have to be collected by hand. Orders can be counted and analysed in whatever level of detail is required.

Techniques of service observation have been developed over many years in the advanced countries. Equipment is available which enables specially trained operators to monitor the setting-up process on a statistically valid sample of calls, and to record success, failure, poor transmission and so on. But manual techniques like these unavoidably consume effort, cost money and introduce the possibility of error. They are also often less satisfactory from the statistical point of view. For example, two operators sampling a fixed number of calls per hour will fail to reflect the variations of traffic through the system over the day. Service in the busiest hours will usually be worse than at slack times. Only computer-based measurement can accurately track such variations at acceptable cost.

ITU requirements

The ITU (see Chapter 13) maintains and publishes world-wide statistics about the most important features of telecommunications in its member countries. These statistics are an important source document for all kinds of agencies and interests throughout the world, including assistance agencies, suppliers and administrations and universities. It is important that each enterprise maintains records which will enable the ITU requirements to be met. A list of the statistics published by the ITU is given in Appendix 3.

Subordinate unit controls

It is central to the working of enterprises which use structures based on B and C in Chapter 3 that they have a proper system of controls for subordinate units. In the most advanced form the general managers' units would be treated as full profit centres, with their own profit and loss accounts and balance sheets.The accounting system must of course be designed or expanded to provide the necessary information.

This approach conforms to the treatment of subordinate units in many modern businesses in the advanced countries, including BT. It is likely to cause the units' managers to place emphasis on the profitability and financial viability of their operations, assessed as though they were independent companies in their own right. It is important that the parallel importance of quality of service is not overlooked. It is all too easy, for example, to endanger quality of service two or three years ahead by concentrating on immediate cost reduction and profitability. For these reasons it is essential in such structures to maintain a system of statistical controls of quality and similar non-financial performance alongside the financial controls.

In cases where the full profit centre approach is not adopted these considerations still apply. Proper statistical and financial and quality controls on subordinate units will still be needed.

Monitoring of performance

There is no point in recording financial detail or statistics unless they are used. It is essential that the principal features of the financial and other statistical performance of the enterprise are reviewed regularly by the top direction, and that the staff are aware that this is done. In BPOT a regular Monthly Report containing the key financial and statistical information was submitted to the equivalent of the Managing Board. Similar arrangements operate in BT plc. Mechanisms for using financial and statistical information to secure the answerability of line managers for the performance of their formations at the end of each year are discussed in Chapter 5.

Summary

To sum up:

— A telecommunications enterprise needs an accounting system based on good modern business practice.
— A number of special problems have to be faced in the case of a telecommunications accounting system, having to do with the nature of telecommunications plant and work-force activities.
— Efficient management of a telecommunications enterprise requires that a system of non-financial statistical controls be operated alongside the accounting system. Such controls should, however, be kept to a minimum.
— Both accounting and statistical systems should make the fullest possible use of computing and telecommunications techniques. The cost of introducing such arrangements should be rapidly recovered in improved efficiency of operation.
— If structures based on B or C are adopted, and especially if subordinate field units are treated as profit centres, it is important that statistical controls on quality of service and so on are operated alongside financial disciplines.

5 Plans and planning

The management of a telecommunications enterprise involves the coordination of a very wide range of activities, from the raising of funds to the recruitment of staff. All the activities interact in a complex way. Also, telecommunications is a long lead-time business. Much of the plant takes two years or more from the completion of planning to bringing into service and it is often provided in tranches designed to last a further three or more years. Major programmes of replacement may take a decade or more fully to penetrate the network. Highly skilled staff take several years to train and develop.

Such an enterprise can only be efficiently managed by preparing coordinated plans which cover all the main parameters of operation — human and logistic as well as financial and technical. If they are to serve their purpose properly these plans must extend several years ahead. The purpose of this chapter is to describe an approach to planning and the use of plans which will meet these requirements. It is based on techniques evolved in BPOT during its period of maximum growth.

The need to contribute to plans imposes important disciplines on the various units concerned. The compilation process itself is therefore of direct value in itself.

It is important when reading this chapter to remember the function and limitations of business plans. All plans are based on forecasts of one kind or another — economic, operational or logistic — and it is impossible to produce perfect forecasts. It follows that it will be extremely rare for the detail of the one-year and five-year plans discussed below to be fully realized. The essential function of planning procedures is to guarantee that overall the actions taken in the enterprise are coordinated and regularly reviewed, and to provide a frame of reference within which changes can be made in an informed way as events develop.

The changes in the variables on which plans depend are illustrated in Appendix 4. This Appendix includes graphs illustrating the behaviour of a number of the principal British telecommunications operational variables over the period 1970–88. The fluctuations of growth will be apparent. In each case the major checks to growth were associated with a slowdown in the UK economy and the revivals with economic recoveries.

For a discussion of general issues which arise in telecommunications planning see Little child, 1979.

Planning documents

Two main planning documents are prepared. They are:

— the one-year plan, setting out financial and other plans for the enterprise in detail for one year ahead;
— the five-year or Medium-Term Plan (MTP), dealing in broader terms with financing, investment and manpower and with major network variables like the number of lines and calls.

It was BPOT practice to prepare a longer-term ten-year plan whose primary purpose was to predict trends in the technology and the environment and to make advance assessment of their implications. It is for each country to decide whether such a plan would be appropriate in its circumstances. It is not discussed further here.

Forecasts

Both one-year and five-year plans rest on forecasts of the main operational variables like numbers of lines and calls for the periods to which they relate. These forecasts have great influence and it is very important that they are properly prepared.

In BPOT extensive regression analysis had been carried out over a number of past years to form the basis of these forecasts. This analysis was constantly repeated and improved as new data became available with the passing of time. Its results embodied the relationship between growth in lines and calls, national economic variables like Gross Domestic Product, Service Sector Output, Personal Disposable Income and so on; and internal variables like changes in BPOT charges.

In the preparation of a particular plan, forecasts of the national economic variables for the period concerned were first developed by BPOT's own economists from their own resources and from study of a number of forecasts by independent outside economists. These economic forecasts were approved at senior level in BPOT. In a developing country it may be appropriate to use economic forecasts promulgated by central Government for general planning purposes.

In BPOT these economic forecasts and predictions of likely price changes were applied to the results of the regression analysis. The resulting central one-year and five-year forecasts of lines, calls and so on were then discussed with the field formations (Regions), which had prepared corresponding forecasts for their territories and had had corresponding discussions with Areas. Any differences between Regional and Headquarters forecasts were resolved in discussion (the process is called 'top-down, bottom-up'). The resulting agreed operational forecasts were submitted for final approval at senior level. Once approved they were promulgated to form the basis of the plan.

The importance of securing top-level approval for the basic economic assumptions and operational forecasts used in planning cannot be overstressed. They are formative for the whole planning process.

Objectives

The planning process works best if the plan is built round a single, clearly defined financial objective. In a state-owned enterprise this objective would probably be set by Government. In a private-sector enterprise it will be set by the Board or by the owners according to circumstances. It might be a specified return on capital or a specified percentage profit.

The plan can also of course embody other kinds of objective, although it is important that there should not be too many and that they should not conflict. For example, in a state-owned enterprise Government may stipulate that investment in telecommunications is to correspond to a certain proportion of Gross Domestic Product in the years concerned. In BPOT in the 1970s the expectation of Central Government was that telecommunications investment would represent about 0.7 per cent of Gross Domestic Product .If the enterprise is owned or controlled by the state some objectives may be set by Government for political or macro-economic reasons and may run counter to the commercial interest of the enterprise. For example, in countries with extensive rural areas it may be decided that the extension of basic service to all communities or to chosen communities by a specified date is to be an objective, even if this causes the enterprise to lose money in the short term. It is important for the business health of the enterprise that when such situations arise its higher direction is properly consulted before a decision is reached; that due weight is given to its views; and that consideration is given to how the enterprise is to be restored to a viable financial situation in due course.

In addition to objectives set by Government in this way, the enterprise itself may, of course, set objectives of its own. For example, an enterprise which is required to extend basic service to all communities by a certain date may decide that it wishes to do so while remaining financially viable. Alternatively it may be decided that a primary objective is to be the elimination of a waiting list for service, the attainment of a specified growth in the number of lines in the plan period or the replacement of a specified amount of old plant. In such circumstances it will be an important part of the planning process to test whether these objectives important part of the planning process to test whether these objectives can be achieved, and if so what increases in charges for customers already connected or other policy changes are involved.

Compilation

In BPOT the work of compiling plans was coordinated by a small central staff unit. The work of such units is formative for the success or failure of the whole enterprise. It is important that they have staff of high calibre who are familiar both with planning and computing technique and with the general character of telecommunications. The unit will have to make difficult judgements in the course of its work, for example, in deciding when the whole plan needs to be recast because one or more elements are turning out to be impracticable or when particular inputs need to be challenged. It should be in close touch with the officals concerned with planning in the top direction.

It is of course possible to prepare plans entirely by hand. But this is cumbersome and expensive and it severely reduces the value of the planning process. It is highly desirable for the enterprise to equip itself with a computer programmed to carry out calculations like the derivation of profit and return on capital and to carry out tests of the sensitivity of the plan to changes in important assumptions like forecast growth of GDP or changes in inflation. The use of computing techniques speeds up out of all proportion the preparation of the numerical parts of the plan. The computer can also test a range of variations of the main planning parameters and evaluate the effect of each on the rest of the plan. For example, a five-year plan built round a specified rate of network modernization may prove to require a level of financing or numbers of specialist staff which are unrealistic. A series of computer runs may determine a practical rate.

The one-year plan

The main features of a one-year plan include text stating the principal objectives for the year in each main field of operations, including support functions like personnel and recruitment, the financial budget for the year, and supporting statements of quality and other performance targets of the kind discussed in Chapter 4.

A one-year plan is prepared annually. In BPOT work on the plan began at least six months before the start of the year to which it related. This allowed time for proper preparation and consultation.

The plan is prepared in the Central Headquarters, drawing on inputs from field formations wherever desirable. If a structure on the lines of B or C in Chapter 3 is adopted, the field formations may be required to make their own proposals for budgets and for performance targets, which may be set in terms of the non-financial statistics discussed in Chapter 4. Such proposals are then scrutinized, debated and agreed between central line management and the heads of the field formations before they are incorporated in the overall plan for the enterprise. Once the complete plan is approved the subordinate unit budgetary and performance targets arrived at in this way can then be ratified as the basis for the operations for the year of the field units concerned.

So far as Central Headquarters is concerned, each functional Department develops proposals both for its enterprise-wide operational sectors, drawing on the approved field forecasts where necessary, and for its own domestic spending. In BPOT for example the central Operational Programming Department would assemble the combined Regional proposals for investment in exchanges and distribution cables, modifying them if necessary in the light of its own knowledge. It would also prepare estimates of its own spending on its staff and so on for the year.

These departmental inputs are assembled by the central planning unit. As the planning process proceeds it is almost certain that individual departmental inputs will have to be revised to take account of the emerging overall picture. It may be found, for example, that the expenditure plans of spending departments will exceed the investment funds expected to be available; and they will have to be adjusted. Often the assumed timing and yield of price increases is found to be unsatisfactory and they have to be revised. Since they affect the forecasts on which the plan is based, this may require the whole process to be repeated.

The effect is to forge a unified approach throughout the enterprise. The end product is an internally consistent document covering the whole income and expenditure of the enterprise and all its operations in concise form.

The plan is submitted for approval by the top direction of the enterprise. Once approved it serves as the basis of the day-to-day operations of the enterprise as a whole and of each field unit for the year concerned.

Subordinate unit plans

If a structure on the lines of B or C is adopted, the process can be taken one step further. Formal one-year plans can be prepared for the subordinate units. Their structures should correspond to that of the plan for the enterprise as a whole, although obviously some functions like procurement which are conducted centrally will not appear in the plans of field units. The subordinate unit plans should be consistent with the overall one. They are prepared in the field units and submitted for approval by the higher direction of the enterprise after scrutiny by the appropriate specialist functions like finance and engineering in the Central Headquarters. Once the plan for his unit has been approved, it is the responsibility of each field general manager to conduct the operations of his unit during the year in accordance with it.

Answerability of field formations

Whether or not separate field unit plans are prepared in this way, at the end of the year the field general managers should be made answerable for the performance of their units against the budgets and targets set for them at the start of the year. In BPOT, formal reviews of performance were held as soon as results for each year were available, when the field line managers answered to their superiors for their results.

It is important that a system of answerability of this kind is supported by appropriate rewards for success or failure. The simplest is perhaps to make the general managers aware that their promotion prospects will be directly affected by the results they achieve. But the management system of the enterprise may also provide for successful general managers to be rewarded immediately by an increase in salary or a single bonus payment (see also Chapter 8); and for unsuccessful managers to be transferred, demoted or in extreme cases dismissed.

The five-year or Medium-Term Plan

A five-year or Medium Term Plan (MTP) has two main functions:

— to provide an approved and coordinated framework within which detailed planning of capital projects and such matters as the recruitment and training of technical staff can be carried out;
— to provide the main input to discussions within the enterprise and with outside agencies like Government and the World Bank about financing and other plans for the enterprise.

Like the one-year plan, the MTP is prepared annually. Each year's plan embodies the last four years of the last plan updated and plans for one new year ('Rolling Planning'). For example, the plan of which 1988 was the first year would have embodied revised plans based on the last plan for 1988, 1989, 1990 and 1991 and a new plan for 1992. The process of preparation may need to start before that of the one-year plan. The planning unit will need to make sure that it keeps the two plans in step, so that the first-year proposals in the MTP are consistent with those in the one-year plan. This is not always easy to do and it requires careful attention.

The procedure followed to compile the MTP in BPOT was as follows:

(1) Forecasts were prepared for all the main operational variables on the lines discussed earlier.

(2) These forecasts were promulgated to the field units (Regions and by them to Areas) to enable them to compile bids for investment in their territories in the plant categories with which they were concerned.

(3) Each Headquarters Department with an interest — network plant, buildings, motor transport, recruitment, training, and so on — was invited to prepare a summary plan for its sector for the MTP period, drawing on the approved forecasts and the bids in (2).

(4) Forecasts were prepared by the finance function of current account (revenue) expenditure and income for the five-year period. This was done by extrapolation of past trends with allowance for known changes due, for example, to manpower savings from the introduction of new technology, to the introduction of new services and facilities or to expected price increases.

(5) All these inputs were absorbed into a single plan and their financial, manpower and other consequences were assessed and summarized. The first trial assembly of the plan could be expected to show unacceptable features. If it did the process would be repeated as necessary until an acceptable overall picture emerged.

(6) Once the main plan was complete sensitivity tests would be carried out on it. A version of the plan might be prepared, for example, based on alternative assumptions about inflation, growth of GDP and so on. The financial and other results of this version would be presented to enable the reader to assess the dependence of the plan on the forward performance of the national economy. Any input variable could of course be tested. In the circumstances of a country short of investment funds, for example, it would be possible to demonstrate the implications of a specified investment shortfall for waiting lists and the general financial performance and profitability of the enterprise.

Submission and uses of the MTP

In BPOT the MTP was submitted when complete for formal scrutiny, modification and approval by the top direction of the enterprise (the Board). If such a plan is prepared, once it is approved it can be used as an input where required to Central Government planning. If the enterprise is state-owned its top direction must expect to have to explain and justify the plan at this stage, and it may have to accept that it is modified in the light of national priorities for capital, foreign exchange, skilled manpower and so on. If such changes are required it is important that a revised version of the complete plan embodying them is prepared and approved.

Approval of the plan also constitutes authority to managers to proceed with those elements requiring action in the first year. For example, because of the long lead-time nature of telecommunications capital plant, orders will require to be committed in Year 1 (say 1989) which will not incur any expenditure until Year 2 or 3 (say 1990 or 1991). In BPOT, inclusion of the relevant expenditure, manpower and so on for a project in the MTP was a necessary but not a sufficient condition for the project to go ahead. It did not constitute authority for expenditure on the project as such to proceed or for contracts to be let in respect of it. This involved quite separate project appraisal procedures and formal approval for the project in its own right (see Chapter 12).

It is important that detailed activity throughout the enterprise is kept in step with the MTP. For example, the estimates of the number of exchange lines used by engineers to plan individual exchange projects in detail should be consistent with the global predictions of the number of exchange lines in the MTP itself. This requires dialogue between those responsible for the design of projects and those responsible for compilation of the global forecasts in the MTP (see Chapter 12).

If steps are taken to ensure that the operational features of the MTP are consistent in this way with detailed planning on the ground, it can be used with confidence to estimate forward requirements for all kinds of purposes. For example, it can be used to derive staff requirements in the different skill groups. From these in turn estimates can be made of the number of staff of each kind which it will be necessary to recruit and train. Further estimates can then be made of how far training requirements can be met by the enterprise itself, and how far it will be necessary to send staff outside the enterprise for training. For training within the enterprise, it will also be possible to determine the numbers of instructors who will have to be found and trained, the building space necessary for training and so on.

In some cases — for example, where general managers are responsible for very large subordinate field units like a capital city's telecommunications — they may find it of value to compile their own MTP. But it is not generally practical to compile MTPs for average-sized field units. Even where MTPs are prepared for subordinate units, it is not appropriate to use them as vehicles for the setting of objectives and for answerability of line managers as in the case of one-year plans. The managers concerned may change once or more during a five-year period. Again, in such a period outside factors like changes in the national economic environment may affect the major features of the plan so markedly that comparison of the achievement in Year 5 of a plan, when that year is actually reached, with the achievement planned for five or six years earlier could be unfair.

Sector plans

It is often desirable to prepare detailed plans for change and improvement in particular sectors of operations. For example, if the priorities suggested in Chapter 7 are observed, it is likely to be desirable to prepare a detailed plan for maintenance and improvement of existing plant. The detailed compilation of such plans is usually a technical matter, on which enterprises may need to seek help, for example, from the ITU Technical Cooperation Department. It is important that such plans are kept in step with the MTP in force when they are prepared.

Summary

To sum up this chapter
— British experience is that properly compiled one-year and five-year (Medium-Term) plans are essential for the efficient conduct of a telecommunications enterprise. A proper five-year plan is also likely to be important as a basis for effective dialogue with central Government, with agencies like the World Bank and, where the enterprise is partly or wholly privately owned, with owners and shareholders.
— One-year plans prepared for subordinate units in Structures B and C provide an efficient way of focusing the answerability of the units' managers for their performance.
— Five-year plans set all major internal parameters like growth, capital requirements and so on alongside external parameters like capital availability and allow the future viability of the enterprise and of its programme of projects to be assessed.
— The process of compilation of plans should use computers to the greatest possible extent.
— Project design and work generally throughout the enterprise should be kept in step with the operative MTP.
— Detailed sector plans may be needed for such matters as plant maintenance. They too should be properly coordinated with the MTP.

6 Charging and pricing

This chapter discusses charging and pricing for telecommunications. It examines a number of issues to do with the pricing of conventional voice (telephone) services. It also considers charging matters to do with the pricing of non-voice (text, data, etc.) services and suggests matters for future review in charging policy.

One general point should be made at the start. BPOT experience has underlined the need for rental and call charging arrangements to be simple and easily understood by customers. A charging arrangement which incorporates too many different conditions, options or variables is likely positively to deter users because they cannot understand what it will cost them to use the service.

Telephone charging structures

In most countries telephone service is traditionally subdivided for charging purposes into:

— rental, covering capital and maintenance charges for the customer's line to the exchange; those (limited) parts of the exchange provided exclusively for him; and the apparatus on his premises where this is provided by the enterprise on rental (see Chapter 14);
— call charges, covering dialled local, domestic long-distance ('trunk' or 'toll'), international and corresponding operator calls.

In Britain a separate lump-sum charge is made for connection to the network. It is designed nominally to cover the irrecoverable cost of connecting a particular customer. This is mainly labour and administrative cost. Virtually all plant except the last length of wiring or cabling into, and the wiring within, the premises can be reused for other customers, and its cost is not treated as irrecoverable.

It was the practice for many years in the United States (and indeed in Britain up to 1929) to combine rental and a fixed sum in respect of local call charges in a single 'flat rate' regular charge. The effect was to make the cost for local calls the same however many were made, and so to stimulate the use of the telephone. The disadvantage was that the 'flat-rate' charge of course had to be higher than the rental would have been. As a result it probably discouraged growth in lines. Also, it is almost certain that a time will come when measured charges for local calls have to be introduced, if only because usage has come to vary widely between users. By then it may not be easy to 'unbundle' charges in this way from the point of view of public opinion. A flat-rate tariff might nevertheless be right for a developing country in its early years.

It is of course possible to offer a choice of flat-rate and unbundled tariffs, provided all the appropriate costs are recovered.

The BT view has been for a long time that flat-rate charges were unfair and undesirable. The general principle of call charging in BT is that rental and call charges should be distinct, and that each call should be charged according to its individual occupancy of the system. This is measured in terms of the distance covered and so broadly the amount of plant used, and the time or duration for which it is in use.

In accordance with these principles local-call timing was introduced in Britain in 1958, and is nowadays taken for granted by customers. Trunk calls have always been timed in Britain as in all countries. At present in Britain all automatic calls are charged for by recording pulses of stated money value on a meter or an equivalent electronic store for each line. The intervals between pulses vary according to the distance over which the call is made and the time of day. The technique is called 'periodic metering'.

Traditionally, local calls have been untimed in many countries. An important advantage of timing is that timed calls may be charged at different rates at different times of day. This enables traffic to be stimulated by load-factor tariffs, which apply lower charges in off-peak periods, when network load is light. Local calls represent such an important element of network load that it is valuable from the economic point of view to be able to apply load-factor tariffs to them.

Pricing considerations

The level and structure of prices for service can only be decided in the light of circumstances in the country concerned. But there are some general considerations that should be mentioned.

The fundamentals of telephone pricing are no different from those of any other business. If a service is provided on a monopoly basis, it becomes a question of policy how prices should be fixed and in particular what element of profit should be included. For example, monopoly rentals can be set by reference to a target return on the capital employed in the exchange and local cable plant and customer premise apparatus if appropriate. If a financial objective has been specified for the enterprise in the form of a return on capital on the lines discussed in Chapter 5, this target return may of course be applied to the individual charging sectors. An important advantage of such cost-based prices is that depending on the elasticity of demand to price they should in theory ensure that demand is at a correct level relative to the rest of the economy. If a service is underpriced it will attract uneconomic demand and consume more than its fair share of national resources including capital. If it is overpriced the service will be insufficiently used and the economy will again suffer.

If a service is provided in competition, its price will be set by the market. It is likely of course that over time effective competition will cause prices to reflect costs, and so stimulate the operators to seek ways of reducing these.

In advanced countries the existing pattern of prices for telecommunications carrier services reflects history. In many of them there have been strong political pressures over the years to hold down the price of telephone service to residential users, who have tended to be most concerned about residential rentals and the price of local calls. These have therefore remained low, with low or sometimes negative profit margins. Long-distance call charges have tended to remain relatively high even though the most marked reductions in operating cost due to the technology have been in long-distance transmission costs.

An unbalanced situation of this kind has certain advantages. Depending on price elasticity in the country concerned, low rentals should stimulate network growth in lines. Calls will not be made unless lines exist and in the long run it may turn out to have been healthy to stimulate growth in lines, even though in the early stages rentals may not have recovered costs or earned a fair return. But as already noted, with time the imbalance will distort the position of telecommunications within the economy and is generally undesirable for this reason. Also from the point of view of the enterprise it is liable to stimulate unwanted and embarrassing demand and affect overall financial performance.

In Britain BT plc inherited a notably unbalanced structure of this kind. At the time of privatization it was clear that the imbalance needed to be rectified; but it was necessary to regulate the prices to be charged by a private sector near-monopoly. When the regulatory regime was devised, care was therefore taken to ensure that BT would not make unduly large increases in charges, either in underpriced sectors or generally in its monopoly operations. For the period up to 31 July 1989, BT's licence limits the increases it may make on its rentals and its charges for dialled calls. In any one year of the five beginning on 1 August 1984, BT is required by its licence to keep percentage increases in the aggregate of rentals and dialled call charges 3 per cent or more below retail price inflation in the year ending on the preceding 30 June. For the five years beginning on 1 August 1989 the corresponding requirement will be to hold prices 4.5 per cent or more below inflation (the arrangement is called RPI-3 or 4.5).

A similar imbalance developed over many years in the United States. Up to 1984 telephone service over most of the United States had been provided by operating companies which were wholly owned subsidiaries of AT&T (the Bell system). There were several other large independent companies and a large number of smaller ones. All these companies held local monopoly franchises and were subject to regulation by the Federal Communications Commission (FCC) and by state regulatory authorities. AT&T provided long-distance facilities to link the local operations (see Chapter 3).

Pressures within the US regulatory system have tended to restrain prices to residential customers in much the same way as Britain. Trends in the technology have been similar. Geographical distances are so much greater in the United States that savings in long-distance transmission costs have been even more marked, yet long-distance call prices have remained relatively high.

In both the United Kingdom and the United States long-haul carrier competition has now been introduced. It has made a situation in which local calls and rentals are subsidized by trunk calls seriously unsatisfactory. The imbalance is being eliminated on both sides of the Atlantic.

To sum up, the correct overall pricing objective from the point of view of management of the enterprise will usually be to adjust tariffs for monopoly services in such a way that each sector — rentals, local calls, trunk calls and international calls — earns a fair return on the capital it employs. If Government decides that this should not be done it is important that this is recognized as a political decision, and not one for which the enterprise must answer.

Demand management and the connection charge

The force of these arguments depends partly on the assumption that the price elasticity of demand for service is high enough to be signficant. BPOT experience has been that at the price levels historically charged in the United Kingdom demand for telecommunications lines is not very elastic to changes in rentals or call charges. Demand usually holds up in face of increases in these.

Experience has shown, however, that in the United Kingdom and in the short term — say over twelve months — demand for new lines is sensitive to the level of the connection charge. In BPOT this charge was used as a regulator. All telecommunications plant takes time to provide. If demand exceeds the levels provided for in the plant capacity on the ground, waiting lists are inevitable. When, as frequently happened, short-term demand rose above or fell below the planned level, BPOT had the option of increasing the connection charge or, if it was desired to stimulate demand, of restraining increase in the connection charge to a level below that justified by costs (circumstances were such that price decreases were rare). This could be expected to go some way to restore demand to planned levels in the short term.

BPOT experience suggests that under conditions of high growth and therefore demand a connection charge should be levied, partly as a short-term means of managing fluctuations in demand. It should be vigorously used as a regulator if circumstances require. If necessary it may be used selectively only in geographic areas of high or unduly low demand. It is in no one's interest for there to be a large unsatisfied demand for service on the one hand, or large surpluses of unused plant on the other.

Future developments in charging

The pattern of telephone call charges used in most countries in the world at present was devised back in the days of manual exchanges. Apart from changes associated with the introduction of customer dialling of long-distance calls, this traditional pattern matched conditions in automatic analogue networks well enough. But traditional charging principles need to be reviewed in the light of the advances in the technology.

A number of countries including Britain are now creating Integrated Digital Networks (IDNs — see Chapter 10). These incorporate stored programme (computer) controlled exchanges which can provide a very wide variety of charging arrangements. Also, these networks will carry all kinds of voice and non-voice (text or data) traffic. The techniques for setting up two or three conversations or their computer equivalent over one channel (2B+D ISDN and 16 kbit/s speech techniques) described in Chapter 10 will make it possible to set up simultaneous voice and non-voice calls through public digital networks. These calls will require their own charging arrangements. It is also desirable to consider the charging requirements and general promotion of the new non-voice services like data and text transmission and 'telematics' (see Chapters 14 and 15). There is therefore a strong case for a review of charging arrangements in most countries at the present time. Developing countries should be in a better position to carry out and give effect to such a review than countries with more developed systems. Some of the questions to be considered are suggested in following paragraphs.

Distance-related charges

There are strong reasons for arguing that the distance term should be progressively eliminated from charging formulae for voice as well as non-voice. Link costs as such must continue to reduce as transmission technology advances. Also, as we note in Chapter 10, in the new processor-controlled networks there will no longer necessarily be a fixed relationship between the plant used to convey calls and their origins and destinations. The physical routing of calls will vary from minute to minute according to network load or failure. The idea of a fixed identifiable correlation between plant cost and call destination is being rapidly overtaken.

Volume-related charges

The BT system for charging for telephone calls was outlined earlier. Telex calls are charged like telephone calls. Packet-switching transactions (see Chapter 10) are charged on a basis related to duration and volume expressed as the number of bits (actually the number of octets of 512 bits) transported per second, with no allowance for distance.

There are certain important differences between voice and non-voice traffic. Most voice calls last two minutes or more. Some non-voice transactions last only tens of seconds. It has been estimated that in a human conversation information is exchanged at a rate roughly equivalent to 50 bit/s. In the slowest data transmission transaction it is exchanged six times faster than this — at 300 bit/s. Over the IDN it will be possible to exchange information without any special arrangements 1,280 times faster than speech at 64 kbit/s. A large amount of text or data can therefore be transmittted in a short time. Put another way, the amount of intelligence conveyed in a given period on a non-voice call at 64 kbit/s is much greater than in a voice call — which in future is also likely to use a 64 kbit/s channel — in the same period.

It is for reasons such as these that the case for a change in the philosophy of telecommunications charging in its bearing on non-voice traffic has been debated for many years. A modern view is summarized in *The Information Network System* by Dr Kitahara of Nippon Telegraph and Telephone Public Corporation (NTT) (Kitahara, 1983). In this he argues that in future tariffs for non-voice services should be based on the amount of transmitted information measured as the product of time and transmission speed (bit rate) with no allowance for distance. BT's packet tariffs reflect similar principles.

The arguments as regards this feature of charging arrangements on the IDN and the ISDN are closely balanced. On the one hand, most present non-voice transactions are carried out at bit rates well below 64 kbit/s which is the rate primarily used for voice. On the other, as we have seen, a non-voice transaction at 64 kbit/s transfers intelligence 1,280 times faster than voice. In practice the IDN will be built up of circuits engineered primarily to carry voice — that is of 64 kbit/s circuits. The use made of such a homogeneous network should in logic be measured simply in terms of the time during which circuits are occupied. The volume of intelligence may vary, but there is no automatic reason why charges should vary with what is conveyed.

It has already been argued that charges need to be as simple and intelligible as they can be. It may be easy for specialists in data transmission to come to terms with complex concepts like charges which vary both with volume and duration. Such experts display great ingenuity in exploiting charging patterns and techniques like multiplexing (Chapter 10) to save money. But such complex arrangements will not make it easy for the main mass of customers to grasp the use of the network for non-voice applications like the mass VANS discussed in Chapter 15. It would be difficult for most people to understand a tariff which varied both in volume and duration. IDN and ISDN transaction charges need to be kept simple. They should therefore be identical as between voice and non-voice, and should vary with duration, but not with distance or volume of bits.

Precision of charging

Another consideration concerns the degree of precision with which charges are applied to calls. For example, in the present British periodic metering system, people who make local calls of, say, half a minute pay just as much as people who make calls of anything up to three minutes, depending on time of day. This is partly because of an inherited expectation that the unit value in pence should be so chosen that the time allowed for one unit on a local call in the most expensive period of the day should range between one and three minutes. Such assumptions and compromises are a traditional feature of charging structures in most countries, and they are accepted. But theory and equity would indicate a structure with a much lower unit value and much shorter charging intervals. This would relate charges much more closely to consumption of network time.

The present arrangements in Britain were largely governed by earlier technical considerations. Charges were imposed by meter pulses applied to the customer's meter by electro-mechanical equipment as the call proceeded. This equipment could not handle pulses faster than about one every second. To accommodate the full range of charge steps, from maximum international to cheap rate local, and to cater for the inherited assumptions referred to above it was necessary to have a wide range of pulse rates. They ranged at times from one every one and one-sixth seconds to one every fifteen minutes.

In some countries (for example, the United States) charges have always been applied to customer-dialled long-distance calls in a different way. At the time of the call the equipment has simply recorded its duration and destination. This information has then been processed separately later to calculate the monetary charge for the call. Such arrangements were avoided in Britain when customer trunk dialling was introduced because of the cost in an electro-mechanical environment. But on SPC exchanges with computerized billing these costs do not arise. As SPC systems are introduced such charging systems will supersede electro-mechanical periodic metering where this has been used. There will thus be no technical constraint equivalent to the speed of periodic meter pulsing. This will mean that on periodic metering systems the cost of all calls can be very closely tailored to their duration by using short charging intervals with a low monetary value per unit.

Separation of setting-up and conversation charges

Another point arises directly from the nature of the hardware. In a processor-controlled digital exchange a large part of the equipment is required only while a call is being set up. This is the part concerned with receiving information from the customer about the destination of the call, processing it, and using it to set the call up. Once this has been done, this part of the equipment can be released and used to set up other calls. Only the part of the exchange concerned with maintaining the conversation path stays in use throughout the call. In traditional charging, no distinction is made between the two parts. But short calls occupy the setting-up equipment for just as long a time as long calls. In this case theory and equity would levy a similar charge for the setting-up process on all calls and differential charges for the transmission path according to duration.

Separation of 'go' and 'return' paths

The next consideration involves another difference between voice and non-voice traffic not so far mentioned. Virtually all voice communications must of their nature be two-way throughout a call. A good deal of non-voice traffic, on the other hand, is (and possibly more could be) one-way for most of the time. Computers do not need to 'interrupt' one another. Control signals may have to be exchanged at the start and finish of a data call, but on many of them the whole flow is in one direction. From the technical point of view it would be possible in theory to separate the 'go' and 'return' paths of all traffic, and to charge only for 'go' while only 'go' was in use. As matters stand, the switched 64 kbit/s channels through the IDN will be engineered to be two-way throughout. It may be right to consider whether they could and should be re-engineered to be one-way; and whether charging and other arrangements should be adjusted so that return paths were only taken into use and charged for when needed.

All charging arrangements require close study related to the circumstances of individual countries and their networks. In practice much of any engineering development involved is likely to be done by suppliers, and patterns of charging are likely to follow largely international models. But developing country enterprises will need to give thought to such matters if they are to know what to ask for and to understand what they may be offered. Changes in charging patterns can only take place at a pace and in a way users can understand and accommodate. The patterns followed in the next few years will have effect well into the twenty-first century. It is important that they should be reviewed now while modern networks are still in a formative stage.

Presentation of call charges on bills

Call charges can be presented on customer bills in different ways, depending on the technical arrangements for metering calls. As we have seen, when customer trunk dialling was introduced by BPOT in 1958, the view was taken that costs should be held to a minimum. No equipment was therefore provided to record the cost of individual calls, except those made through the operator. As will be noted from previous paragraphs, the metering system recorded units for all dialled calls on a single meter. The bill shows only the total of call units and the charge for it. There has been sustained consumer pressure over the years for bills which would list dialled trunk and international calls in detail, showing destination, duration and so on. On SPC exchanges it is possible to arrange this without significant extra engineering cost and BT has plans to do so. Itemized bills of this kind are standard in the United States and many other countries.

Summary

To sum up:

— The general principles of charging in Britain are that the charge should be separated into rental, connection charge and call charges; and that all calls including local calls should be charged according to their individual occupancy of the network, in terms of time and distance.
— BT pricing principles are that competitive prices should be set by the market, and that where no competition exists each sector — rentals, local, long-distance and international calls and so on — should earn a fair return on the capital employed in it.
— In the United Kingdom price elasticity of demand in relation to rental and call charge increases is low; but in relation to a lump sum connection charge it is relatively high in the short term. In such circumstances the connection charge can be used as a short-term regulator of demand for lines if this is needed.
— Call charging patterns should be reviewed in the light of advances in the technology and with the particular objective of promoting non-voice traffic (data and telematics — see Chapter 15).

7 Priorities and internally generated capital

Telecommunications is highly capital-intensive, and its expansion depends on adequate investment. The problems of arranging such investment in developing countries were central to the work of the Maitland Commission. Sir Donald Maitland refers in the Foreword to recent ITU publications which discuss the issues in detail. These problems are linked to questions of priorities which face administrations in most countries. This chapter examines these issues in the light of the Commission's conclusions.

The reader's attention is drawn to paragraphs 1–8 of Chapter 9 of the Commission's Report and in particular to Paragraphs 9 and 10. These say:

9. ACCORDINGLY WE RECOMMEND that developing countries review their development plans to ensure that sufficient priority is given to investment in telecommunications

10. WE FURTHER RECOMMEND that developing countries make appropriate provision for telecommunications in all projects for economic or social advance and include in their submission a checklist showing that such provision is being made.

It is assumed in this chapter that these recommendations have been accepted and are being acted upon.

Priorities

The theme of earlier chapters is that telecommunications deserves to be treated as a modern business and to be managed by modern management techniques. Also that the experience of advanced countries and Britain in particular makes it possible to suggest how such techniques can be applied to effect, in a manner suited to the special characteristics of telecommunications. It is, however, important to recognize that in many countries the administration wishes to improve its telecommunications but starts from a very difficult position. An earlier administration or a commercial operator may have left behind a legacy of obsolete and poorly maintained plant; there may be severe shortages of lines and of capacity for calls; large tracts of country and numerous centres may be wholly without telecommunications; capital and in particular foreign currency may be very scarce; and trained staff very few in number. Costs are likely to be high and service can be poor. In these circumstances it may be helpful to consider the priorities for action, drawing on BT experience. Each country has its own set of problems and must make its own decisions on priorities. But certain principles can be suggested.

Telecommunications is not just a business enterprise but a vital public utility. It is suggested that the first priority must be to give an adequate internal and international service to those sectors of the community which have a central role in public administration and the national economy. This is likely to include not only Government and public services but major business enterprises which earn foreign currency or generate substantial employment or both. It is suggested that such capital as is available should be used first for this purpose.

The second suggested priority is to set in place a modern management structure, with an adequate accounting system and other information and control procedures, on the lines discussed in this book. Only when this has been done will it be possible to ensure that the enterprise is doing all it can to help itself, that investment in new plant is correctly deployed and that money is not being wasted. And once it has been done the enterprise will be in a much stronger position to make its case to central Government for new investment, or to attract support from agencies like the World Bank.

In a country which faces a shortage of trained staff, as many do, it is suggested that training should come next. It is essential to create as soon as possible a cadre of management and professional engineering and computing staff of graduate calibre, and to train the necessary force of technicians and operators if these are not already available.

The fourth suggested priority is to achieve the best service possible for the remainder of those already on the telephone. This is not only because telecommunications is a vital feature of an advancing society, but also because it is essential to maximize income and so enable the enterprise to help itself financially to the greatest possible extent. Plant which is idle because it is defective or not properly connected into the network does not earn money. A great deal can be achieved with old obsolete plant if it is properly exploited and maintained. BT still has a number of exchanges giving tolerable service — though at a cost — which were installed in the 1930s and went through the Second World War; and parts of the BT distribution cable network also date from the 1930s.

Finally, the Maitland Commission laid stress on the importance of extending at least minimal telecommunications service to every community, however remote. Radio and satellite systems (Chapter 12) make this less prohibitively expensive than it used to be. From the national point of view it deserves priority attention.

It is suggested that these steps should be set in hand before the enterprise addresses any major modernization or expansion of its network. There is no point in adding lines if the service is so bad that they are largely useless. And it is important to bear in mind that the enterprise will have to live with what is done in its network for many years to come. It has the most powerful motive to get it right. External advisers, suppliers and consultants can make a big contribution in terms of skills and know-how. But the management of the enterprise should aim to stand on its own feet and take its own decisions on plans, procurement and so on as soon as it can. If it does not have staff with the necessary skills, it should create them as soon as possible.

Determination of investment requirements

Once proper planning procedures are in place it is possible to address the investment problem on a sound basis. It is a central function of the planning processes described in Chapter 5 to determine the investment programme of the enterprise. If a five-year plan is prepared on the lines described it provides a yardstick for what is needed and a framework within which projects can be identified and worked up. If projects are conceived in isolation from the general capabilities and financial position of the enterprise there is a danger that it may be found later that the enterprise cannot support them. A project which cannot be supported in terms of the ability to maintain the plant, load it with traffic, meet interest and depreciation charges on it, or staff it can be a serious hindrance. The plan will also give a solid basis for discussions about investment with central Government and agencies like the World Bank.

From the point of view of the enterprise the best measure of a profile of capital requirements is therefore a validated five-year plan. Discussions with Government and other outside agencies will be helped, however, if there is an objective measure of the absolute minimum investment which is justified from the point of view of central Government planners. In circumstances of real shortage of capital and foreign exchange it may not be easy to derive one. If the priorities suggested above are accepted, however, it is possible to suggest an approach to determine requirements for service to the key elements of the economy and the State.

Recent research in BT has quantified the penetration of telephone service into business sectors in the United Kingdom. The results are set out in Figures 7.1 and 7.2. These are figures for an advanced economy with developed telecommunications; and even then it will be seen that penetration was still evolving in Britain as between 1982 and 1987. It should therefore be stressed that these results can only give a broad indication so far as developing countries are concerned, but they should still provide a useful yardstick.

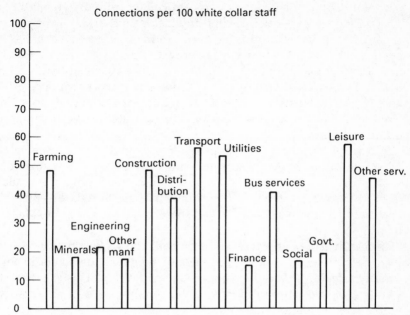

Figure 7.1 Telephone intensity by sector, December 1982

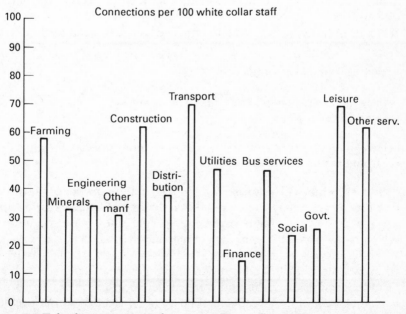

Figure 7.2 Telephone intensity by sector, December 1987

General national development plans should indicate the sectors among those shown in the figures to which it is desired to give priority. It may be part of a national plan, for example, to give priority to agriculture and primary industries, to construction and to transport. It should be possible for most developing countries to derive from experience an average cost per line of the investment needed to provide service to such undertakings from existing plant; and to apply this to figures for telephone requirements derived for the country concerned from past experience on the lines of the BT statistics in the figures. In this way it should be possible to build up an immediate minimum telecommunications investment profile geared to national priorities.

It is emphasized that this approach is designed only to ensure that telecommunications gets a bare minimum of investment rather than none at all, in an economy which is really short of funds. If funds are available from any source for major development projects it is essential that telecommunications gets a balanced share, as advocated in the Maitland Report.

In circumstances of severe shortage of plant it is likely that service will have to be refused to large numbers of potential customers in lower priority categories. Large waiting lists may develop. This is bad for the standing of telecommunications in the community; and an enterprise with a bad reputation may find it that much harder to get its share of investment. Also, the administration of waiting lists is a difficult, time-consuming and often invidious business. Steps can be taken to contain waiting lists by pricing measures. In particular, it is suggested in Chapter 6 that a lump-sum connection charge may be used as a regulator of demand. Such payments by customers are themselves a source of revenue and therefore of investment. They are discussed further below.

Internally generated capital

As has just been suggested, funds to meet investment requirements can be found in two ways:

(a) by generating them internally within the enterprise;
(b) from external sources.

Internal funds will of course be generated in the currency of the country concerned. They may therefore be of only limited value as far as major projects involving foreign suppliers are concerned. But an enterprise which has adopted the approach to management advocated in this book will be equipped to do a great deal to improve matters by less ambitious measures. Very worthwhile programmes for overhaul and refurbishment of older exchanges, radio and cable systems or for connection of new customers will often be capable of being mounted primarily using local currency.

If the enterprise operates an accounting system on the lines advocated in Chapter 4, funds for investment will be generated internally from provision for depreciation and from operating surpluses (assuming prices and volume of business allow such provision and surpluses in the first place). The level of funding generated by depreciation is partly a matter of accounting practice applied to the asset structure of the enterprise and partly a matter of policy. For example, many enterprises have inherited technologically obsolete plant which is likely to have been depreciated over long periods — twenty years or more is not unusual — and to be largely written off, with only a low residual value. If such plant is treated along orthodox lines it may not warrant much conventional depreciation provision. Yet management may be aware that it is urgent to spend large amounts on new plant, greatly in excess of the conventional provision being made.

BPOT faced this situation in 1955. It was decided then to make additional ('supplementary') depreciation provision according to the best estimate that could be made of a reasonable annual provision for new plant. This supplementary depreciation was treated as a cost for the purposes of price fixing. The effect was to cause existing customers to make a larger contribution than they would otherwise have done to the foreseen cost of replacing the exchanges and plant they were using. From 1961 the amount earmarked in this way counted as a contribution to the profit which BPO telecommunications were required to earn, expressed as a return on capital. This return was defined as Percentage Return on Mean Net Assets (RMNA) — that is ((Profit after (combined Depreciation Provision + Interest Payments on State Loan))/(Mean Value of Assets at Replacement Cost)) × 100.

The financial performance required of BPOT by Government was specified in terms of RMNA defined in this way. This target was reviewed annually in discussions between BPOT and Government as part of the main planning process. These discussions took account of the Self-Financing Ratio (SFR) which would be generated — that is ((Profit + Historic Depreciation Provision + Replacement Depreciation Provision)/(Capital Requirements in Year)) × 100. Throughout the 1970s BPOT was encouraged to achieve high SFRs and was able to do so. This was important because it very considerably reduced BT dependency on Government for investment and therefore increased the management's freedom of action.

It is open to the enterprise to introduce measures designed directly to increase the surpluses available for reinvestment on its own initiative. If a charge is made for connection to the system on the lines advocated above, for example, it may be set at a level which exceeds the cost of connection. Depending on the price elasticity of demand, it will then not only control demand but generate funds. Alternatively, rentals and call charges may be pitched at a significantly higher level than would be justified by costs plus normal profit, again in order to generate funds for investment. It is important of course that information is collected about price elasicities in the country concerned before such steps are taken; otherwise there is an obvious danger that the surcharges may be greater than the market will bear.

The idea of a lump-sum payment on connection can be further developed in various ways. For example, new customers may be required to take out a capital bond before they are connected to the system. Such a bond may earn interest and be subject to repayment at some future date. But it creates a convenient way of generating short-term funds for investment.

Finally, the enterprise can operate a policy of active cash flow enhancement in other ways. Once a proper accounting system is in place, it should be possible to determine the profitability of each sector of operations. It may be found, for example, that customers in city centres or outlying areas make more use of the telephone, and therefore generate greater surpluses, than those in suburban areas. It may be found that provision of some kinds of CPE (see Chapter 14) is more profitable than others, and so on. Once such information is available marketing and publicity effort can be mounted to stimulate business in the areas concerned. An enterprise which manages its affairs actively in this way is clearly likely to earn credit with its own Government, with agencies like the World Bank or with its owners or shareholders if it is in the private sector.

Externally generated funds

However skilful the management of the enterprise it is unlikely that all investment requirements will be met from internally generated funds until very high levels of penetration are achieved. And requirements for foreign exchange cannot generally be met from this source.

External sources of funds are:

— State or other local public loans (which may be in short supply and will usually be in local currency);
— loans of various kinds made to the Government from the World Bank and similar agencies (where serious conflicts of priority may arise between telecommunications and other applications);
— loans from bilateral sources (which will usually be tied to particular projects and/or suppliers and require careful consideration to be sure they meet the overall needs of the enterprise and can be repaid and otherwise supported);
— funds raised by constituting the telecommunications entity in such a way that it can raise money from the private sector (which involve political issues outside the scope of this book).

The first three of these are considered in more detail in the Report of the Maitland Commission.

Summary

To sum up:

— Telecommunications priorities are a matter for each country to decide for itself. But BT experience does allow a set of priorities to be suggested for consideration.
— The capital needs of an enterprise should be assessed globally on a year-by-year basis through proper planning procedures rather than on a project-related basis.
— The best way to assess them is to prepare an annual five-year plan on the lines set out in Chapter 5.
— Recent research in BT has suggested a technique by which bare minimum telecommunications investment requirements geared to general national development priorities may be estimated.
— The proportion of requirements met from internally generated funds can and should be maximized by skilful management of the enterprise. But it is likely to be well short of what is needed, and in particular to fail to meet foreign currency requirements.
— The various external sources of investment each have their own complications. Many of the issues are political. They are a matter for the Government and international agencies concerned in the light of the conclusions of the Maitland Commission.

8 Staff and staffing

Although telecommunications is capital-intensive, its success or failure depends in the end on people. This chapter deals with a number of important matters to do with telecommunications staff and their management.

Staff costs are usually a main, if not the dominant, item of operating expenditure for telecommunications enterprises. While much depends on the employment climate and practices of the country concerned, it is usually easier to recruit staff than to shed them. An enterprise which is expanding rapidly is likely to be recruiting extensively. In such circumstances it is particularly important for line management to maintain close control of staff numbers and to oversee the quality of staff recruited.

Traditionally in many countries telecommunications has been seen as a career job for management, supervisors and skilled technical staff of all ranks. In the past this was true in BPOT. But changes in British employment patterns and in the technology have caused BT to change its approach. It now aims at a balance between career staff and staff on shorter-term contracts. In appropriate circumstances it may be correct for developing countries to adopt a similar approach. This will make it easier to accommodate peaks of work, for example, on network replacement, and to shed inefficient or otherwise undesirable individuals.

Determination of staffing requirements

The planning process described in Chapter 5 involves predictions of future staff requirements in broad categories. In preparing the MTP it would be appropriate to predict separately requirements for management, engineering professionals (graduates or equivalent), basic engineering (technicians and others), operating staff, office (commercial and administrative) and computing and other specialist staff. In making these predictions account will need to be taken of the staff productiveness objectives of the enterprise (see below), of prospective changes in work organization and content — for example, due to technological change — and of training times for each group of staff.

Detailed personnel planning requires predictions in greater detail. Techniques for calculating requirements for this purpose for each of the main skill groups are discussed below.

Requirements for basic engineering staff

Basic engineering staff are the biggest single group. For a more detailed discussion of BPOT philosophies for the determination and efficency improvement of staffing in this group, see Merriman (1975).

If no standards exist requirements for this group may be calculated by techniques which allow in detail for the work to be done. A time unit value, for example, in seconds, can be arrived at by work study for each item of work such as maintaining or installing an exchange switch, printed circuit card or relay set. The results are multiplied by the number of items of each kind and the number of times the process has to be repeated per month or year to arrive at the theoretical workload. But this is a lengthy and laborious process to carry out from scratch.

Most enterprises will have inherited their own staffing levels and practices. The main concern is likely to be how to modify them to improve efficency. New standards may, however, have to be derived for requirements for new equipment like SPC exchanges (See Chapter 9).

Engineering staffing and training programmes need to be planned for several years ahead. This is because the training of fully qualified technical staff takes up to five years; although in the later part of their training they may be employed on the job in support of qualified staff. It is likely to be desirable to plan recruitment and training programmes within the time frame of the MTP, that is for five years ahead.

Special problems arise in planning engineering staffing in undertakings which are embarking on major network expansion or replacement programmes (see Chapter 12). There is a real danger that larger numbers of engineering staff will be recruited for such programmes than will be needed once the programmes are completed. Surplus staff can be a serious embarrassment and can lead to low efficiency and morale. It is important that plans are made with this in mind. In some cases it may even be desirable to slow the rate of expansion or replacement or to employ contractors or staff on short-term contracts to do work normally done by the standing staff of the enterprise to avoid such situations arising.

The economic justification for introducing modern plant like digital exchanges often depends heavily on savings in engineering maintenance staffing. It is important that staffing is controlled to ensure that the predicted savings are actually achieved. BT experience is that this requires close management attention.

Requirements for operating staff

There are very well-established procedures for calculating operating staff requirements.

The present operator work load is first measured by detailed sampling and counting of actual traffic. A prediction is made of likely growth or reduction (for example, because an exchange is to be converted from manual to automatic) and the result is converted first into units of time by applying appropriate standards and then into the staff numbers required to cope with the expected work, with allowance for objectives in staff productiveness.

This may be done annually if conditions are stable. But in many developing countries it is likely that there will be frequent changes in operating load as exchanges are converted from manual to automatic; as computers are introduced into directory enquiry; as customer dialling of trunk calls is introduced; or as traffic grows. The situation needs to be closely monitored to make sure revisions of staffing are undertaken whenever necessary. The economic justification for conversion to automatic operation often depends on operating savings, and once again it is important that these are actually realized.

Effect of automatization

Particular problems may arise with engineering and operating staff due to the conversion of local and long-distance service and exchanges from manual to automatic.

It is likely that some operator switchboards will become surplus and require to be closed down. The planning of arrangements for such situations and for the dispersal of the staff affected requires close attention. It is likely to be particularly difficult to shed large numbers of operating staff immediately on conversion; and it can of course cause hardship for the staff concerned. It is important that plans are made in advance to minimize this problem. It can often be contained, for example, by recruiting temporary staff, by arranging to absorb surplus staff at a nearby exchange where operators are still employed, or by voluntary redundancy.

The composition of operator traffic will be markedly different after conversion. When customers can dial all calls for themselves they will call in the operator only when they need assistance or special facilities like personal or transfer charge calls. Such traffic takes a lot of operator time and is therefore expensive to handle. Its volume can be controlled by appropriate charging arrangements.

Much of this work is of higher quality than the old-fashioned labour of connecting straightforward calls. Nevertheless, BPOT experience is that the morale of the operating staff may suffer from the sense that they no longer occupy a central position in the operation of the service. Such problems are greatly helped if they are allowed to participate in wider services, like answering services, advice to customers on apparatus and services available and so on.

Engineering staff requirements will be affected in a similar way when older exchanges are replaced by electronic exchanges which require much less maintenance. The number of engineering staff affected at any one centre is likely to be less than in the case of operators, but requirements for them also are likely to be markedly reduced. Similar measures to those suggested for operating staff may be needed to contain problems.

Requirements for office staff

Office staff costs are an important element in operating overheads, which can sometimes be neglected. It is important that they are given the attention they deserve.

The staffing of the larger office groups, for example, on customer accounts, deserves to be approached in the same methodical way as engineering or operating staffing. The measurement of the work of office staff presents difficult problems because it is often varied and unpredictable. These problems are not of course confined to telecommunications. For a discussion of techniques for determining office work-load the reader is advised to consult specialist sources. Once the work-load has been calculated, the procedure for determining requirements is similar in principle to that for the other skill groups.

Large sectors of office work are often appropriate for the use of computers. This will usually involve radical change in work processes and systems. The design of substantial computerized systems is a job for specialists. It may be difficult and inappropriate for developing countries to build up large teams of staff concerned with this work. It will often be carried out by outside contractors or consultants. But it is most important that the enterprise has permanent staff of its own who thoroughly understand the computing systems and can control, maintain and modify the software. It is good practice to associate with a project from the outset both eventual users, who can influence the design of the system, and people who will have to assume responsibility for running the system and maintaining it when it is complete.

Computing projects can give rise to similar staffing problems to those associated with exchange conversions. Such problems may arise from redundancy among the staff who did the work before it was computerized or among staff directly associated with the project, like operatives keying new databases. They should be handled in a similar way to those involving operating staff.

Managers and professional staff

The requirement for managers, engineering graduates, computing staff and other specialists can rarely be calculated in a straightforward way. Careful judgements will be needed. Thorough consideration of forward needs for such groups of staff is an important responsibility of the top direction of the enterprise. Assistance from outside bodies like ITU/TCD, the Centre or consultants may be appropriate. BT experience indicates that it is imperative that an adequate body of trained managers, professional engineers and computer specialists is created and maintained.

Recruitment

Once staffing requirements are known recruitment requirements for each group can be calculated. The process should allow for the number of existing staff, for planned changes in requirements and for wastage. (Wastage is the percentage of staff in each group leaving each year due to retirement, death, changing jobs and so on.) Wastage is normally higher among female staff, who leave to get married and have children. The enterprise will have its own policies on treatment of married women and re-employment after child-rearing. Re-employed married women are often very valuable, because of their experience and because they do not need full re-training when they resume work.

Telecommunications enterprises like BT now use advanced techniques for selection of recruits. The most sophisticated procedures are used for graduate managers and engineers, but techniques for selection of all kinds of staff have advanced greatly. Personnel selection is a specialized subject, with an extensive literature.

Training

Telecommunications is a high-technology business, and it will not work without adequate numbers of properly trained staff. The staff determination procedures described above provide the basis for a methodical way of determining the numbers requiring training in each skill group in each year. It is essential that proper arrangements are made actually to train them and that the effectiveness of what is done is kept under close scrutiny by the top direction of the enterprise.

The training needs and methods of a particular enterprise will depend heavily on the geographic and demographic layout of its area, levels of education among the general population, existing facilities in the enterprise and in universities and so on. BT has an extensive network of engineering and other basic staff training schools. This includes a central engineering training school for the most skilled staff and regional centres throughout the country. BT also maintains its own Management Training Centre. It makes use of outside training for individuals with special talents or responsibilities.

BT makes extensive use of computer-based training techniques. These are well suited to the teaching of a wide variety of specialist skills, including engineering for technicians and other basic staff. Computerized training can be carried out over telecommunications links to remote points or centres. It should be well suited to the needs of developing countries. It must be emphasized, however, that computerized training is only effective as part of a course managed by human instructors, who need to be available to be consulted.

Management training needs to include both basic skills like manmanagement, communication and elementary accounting and instruction in the actual work to be done like running a maintenance unit or an accounts office. It may be helpful for enterprises in smaller developing countries to arrange for their managers to be trained abroad. But whenever practical it is suggested that an enterprise should develop its own management training, in conjunction with local universities and other institutions.

The approach to be adopted by developing countries in organizing training was studied by the Maitland Commission. The reader is referred in particular to Paragraphs 6–23 of Chapter 7 of the Report for an examination of the issues that arise.

Staff productiveness*

The productiveness of the staff can have a major effect on the economics of a telecommunications enterprise. It calls for specialist attention in its own right.

Engineering staff productiveness

Most of the questions involved can usefully be addressed in considering the productiveness of engineering staff. (For a more detailed description of BPOT practices in this respect, see Tomlinson, 1983 and Turner, 1971/72.) This depends on three distinct factors:

(a) The labour intensity of the plant — digital exchanges require fewer man-hours to maintain than electro-mechanical exchanges because they are more reliable and because on site maintenance procedures are

---

* The word 'productivity' is often used in this context. It has a wide variety of connotations, however, and in the interest of clarity it is not used here.

simpler. Some of the staff may, however, have to be more highly qualified because the technology is more advanced.

(b) Output per individual — one man may overhaul ten switches where another only does five, simply because he as an individual works faster.

(c) How the staff are caused to use their time — an exchange maintenance man may quadruple his effective output if he is given a better diagnostic tester. A cable gang may treble theirs if they are given a mechanical device for drawing cable through duct; but they may then waste the better part of a day simply because such a device is out of order. All kinds of staff may waste time simply because they are badly organized and slow to start work; or because they are put on the wrong work.

Of these (a) is a matter of technical policy (see Chapter 11). (b) is a matter of man management in the direct sense. It involves questions like how staff are selected, trained and motivated. (c) is a basic function of management.

In the engineering field gains in (c) come in three ways:

1. Better organization of work and staff — the key here is the quality and skilfulness of first- and second-line supervision, and the fostering among them and among their staff of a sense of personal responsibility and accountability.
2. Positive projects, like improved mechanical aids for laying cable or erecting poles or improved ways of organizing routine maintenance work. Such projects may well need to be devised and implemented on a steady programme across the whole area of engineering activity.
3. Close control and oversight of engineering recruitment and of man-hours and overtime to ensure that the benefits of 1 and 2 are not swamped by uncontrolled recruitment or consumption of man-hours. Specialist managers with this responsibility should be appointed in line engineering formations.

BT introduced a programme to improve (c) by measures like 1, 2 and 3 in its engineering operations in the 1960s. Despite this, comparison with the best of AT&T (Bell) performance in the late 1970s suggested that BT still needed 30 per cent more staff to carry out the same work — for example, to maintain a Strowger (step-by-step) exchange of a given size — than were needed in the United States. Developing countries may face similar problems.

A drive to get the most out of (a) and to improve (b) and (c) is an essential element in efficient management of most telecommunications enterprises. But it should be carefully controlled and monitored. An over-ambitious drive to improve productiveness and save staff will cause quality standards to drop and will impair service to customers. It is particularly important that productiveness objectives (or staff shortages) are not allowed to impair engineering workmanship. If they are (or if sloppy work is allowed for any reason), lasting damage will be done, especially in vulnerable and little-visited parts of the network like distribution cable joints.

It is often assumed that productiveness and quality work against one another. This is not so. Good workmanship at the time of installation will both protect service for the future and save maintenance man-hours.

Operating and clerical staff productiveness

A similar approach is appropriate to foster productiveness among operating and clerical staff. Special problems are likely to arise with women staff, whose wastage is usually higher than that of men because they leave to get married. The effect is to impose a special load on training and to dilute the skills of the staff group concerned. Close attention to the quality of training and supervision is needed in these circumstances.

Control of overtime

A typical basic working week is perhaps forty hours, although this varies from country to country. If staff work longer hours than this they are said to work overtime.

Good control of overtime can make an important contribution to productiveness. In managing overtime it is important to remember that a premium rate — say, time and a quarter — is often paid for overtime. It must also be said that excessive levels of overtime — say, over five hours per week — lead to tired staff, illness and loss of efficiency.

There are three main kinds of overtime:

— Unavoidable overtime — this has to be worked by staff who must work through until they finish, for example, repairing major cable faults.
— Overtime to absorb peaks of work — it is often preferable to work controlled overtime rather than to recruit staff to absorb short-term peaks of work. It is important for managers controlling recruitment to learn when to schedule overtime and when to recruit.
— Planned overtime to gain efficient output — the length of the standard day or working week will be set by custom or negotiation. Staff can often work a limited amount of extra overtime while maintaining full efficiency. If the work is there overtime up to about an hour a day — four to five hours a week — will be efficient, provided always that supervision ensures that work is actually done in the extra time.

Measurement of productiveness

There are many ways of measuring staff productiveness. It was found in BPOT that for internal management purposes the most useful measures related physical products (telephones or cable pairs installed or maintained, etc.) to man-hours expended. Thus the standard measure of overall engineering productiveness ('productivity') throughout the period was (Total Engineering Man-hours in 000s (KmH)) divided by (Total Telephone Stations). BPOT experience was that with a compound rate of system growth in lines (connections) over the period 1965–83 of 6 per cent p.a. and of stations (telephones) of 6.5 per cent p.a. it was sensible to look for an annual improvement of 5 per cent in productiveness measured in this way. It was found that this was a demanding rate of improvement, requiring ceaseless vigilance, if quality was also to be preserved. Higher rates might be possible with higher rates of growth, but great care would be needed in trying to attain them.

More refined measures of this type may be useful for management purposes for particular blocks of work. For example, maintenance staff productiveness may be determined by (Maintenance Man-hours in KmH) divided by (Total Stations). Note that this kind of measure is quite different from the work-study-based measures of time per item of equipment discussed above.

To communicate to the staff themselves how good or bad their personal performance is from this point of view it will be useful to publish figures for their performance expressed in variables they can directly influence. For example, engineering maintenance staff can be expected to grasp without difficulty the significance of the number of repeat (that is, recurrent) faults which arise on plant they have serviced. The number of such faults affects both quality and productiveness.

Staff rewards

Staff cooperation is of course an important element in efficiency improvement. If productiveness is measured by relating products to man-hours consumed on the lines suggested above it is possible to foster improvement by including an element related to gains in productiveness of the particular staff group concerned in pay settlements. But it is often preferable to reward staff efforts which have contributed to gains in operating efficiency through mechanisms related to the financial performance of the enterprise. Staff may, for example, be given a bonus deriving from the profit or surplus achieved by the enterprise as a whole or by the unit in which they work if this is a profit centre. In a company which has been wholly or partly privatized staff can be rewarded with shares if this is judged appropriate.

There are two reasons why financial schemes are often better. In the first place, the techniques of measurement described above are satisfactory for internal management purposes. But they are not really precise or robust enough to form the basis of pay settlements; whereas financially based schemes are based on hard fact. Secondly, in most cases gains, say, in engineering are likely to be due in some part to groups of staff other than basic engineers — for example, management. It may be difficult to cater for them in measurements related to basic staff man-hours, but they stand to benefit from overall profit-sharing arrangements.

The decision to introduce incentive payments of any kind is a matter of policy for each enterprise to decide for itself in the light of local circumstances and practices.

Work-force organization

BT experience has indicated some important considerations affecting the organization of the work-force.

Engineering

Field engineering work requires the following groups of staff:

1. Second line managers (typically graduates or experienced promoted staff with the necessary technical knowledge)
2. First line managers (typically people with higher secondary education or promoted technicians with the necessary technical qualifications and potential for management).
3. High-skill 'internal' technical staff, employed in exchanges, transmission repeater and microwave stations and on installation and maintenance of PABXs and similar customer equipment (highest available technical qualifications below graduate level).
4. Other 'internal' technicians, supporting 3.
5. 'External' chargehands — typically in charge of cable installation or pole erection gangs.
6. Skilled 'external' staff employed as cable jointers and on installation and maintenance of customer lines.
7. Semi-skilled 'external' staff employed in cable gangs and on similar work.

(Groups 3–7 will normally receive the bulk of their technical training in or through the enterprise.)

Specialist staff like those on Motor Transport, Building Construction and Maintenance and so on normally work in structures reflecting practice in their professions and trades outside the enterprise.

Each of Groups 1–7 gives rise to its own issues.

It is desirable that as many as possible of Group 1 — second line management — should be fully qualified engineering graduates. In some countries engineering management work like this, or the corresponding engineering work in the central headquarters like network planning, may not have the same prestige as research and development. It may therefore be difficult to recruit and retain a sufficient number of graduates in such work. It is important that high-calibre people recruited to this group are encouraged and helped to appreciate the interest and significance of managerial problems and the promotion prospects for successful engineering managers and planners. Senior management can do much to influence the outlook and standing of individuals at this level through personal contact and in other ways.

The importance of Group 2 — first line management and supervision — cannot be overemphasized. However good the staff at lower levels, they will not work to maximum effect unless they are led by managers who are good leaders and organizers and have a sufficient grasp of the technology.

Group 3 — high-skill technical staff — can also pose important issues. For example, exchange maintenance work on modern systems demands highly intelligent people, with analytical skills and many of the qualities of supervision. Such qualities are likely to be at a premium in most countries. It may be necessary to grant such good pay and status to attract and retain these staff that many of them who would otherwise have made good managers are content to remain at this level and avoid management responsibility; or they may tend to look for similar jobs elsewhere rather than accept promotion. The effect is to deny good potential first line managers to the enterprise.

It is important that the organization and grading structure avoid encouraging the formation of such attitudes. The relative pay levels of senior technical staff and first line managers obviously need to be kept in the correct relationship. Again, the work of maintaining high-technology plant can often be broken down into specialisms, many of which do not call for people with special qualities. A small number of staff of really high calibre will always be required, for example, to maintain and update software and to repair high-technology hardware faults. But they should generally be few in number. This should mean that they have good prospects of promotion and can be encouraged to look for advancement within the enterprise.

It is important that the structure provides for staff to move between groups. For example, staff in Group 6 should have incentives and opportunities to gain technical qualifications and move into Group 4 and later Group 3; and staff in Group 7 should have similar openings to move into Group 5.

Operator services

Under modern conditions the requirements for organization of operating staff are straightforward. The groups involved are:

1. Second line management (typically graduates or experienced first line managers).

2. First line management in day-to-day control of switchboards (usually experienced former operators or supervisors).
3. Supervisors, who directly supervise operators at the switch board and may also handle difficult callers.
4. Operators.

A majority of operators are usually women and, as noted earlier, a relatively high turnover is likely in this group of staff. Training is therefore a continuous business. Operator training may be organized either centrally, at dedicated schools, or locally on working switchboards. BT has used both approaches. In the circumstances of most developing countries local training is perhaps to be preferred, provided centrally trained instructors are available. There may be an exception in large cities and conurbations, where turnover is particularly high and training can be efficiently organized at a central point.

Office and administrative staff

Office and administrative staff are usually organized in a single structure comprising basic staff, supervisors and management and covering a wide range of work. Under modern conditions it will be important in assembling managerial and similar staff to aim for the right mix as between qualified specialists like accountants and computing experts and generalists, and as between people recruited outside and promoted staff. To ensure a good supply of candidates for promotion it is important that junior office staff are given opportunities to broaden their experience and gain advancement. In particular they need the chance to understand and gain familiarity with telecommunications and computing technology, so that they do not feel themselves disavantaged as compared with engineers and other specialists.

General staff management considerations

As a general matter the structure and personnel practices should foster qualities of leadership among junior staff who may be potential future managers and give them appropriate opportunities for advancement. It is good practice to prepare regular reports on the performance, qualities and potential of all staff.

It is important that all middle and junior managers are led to appreciate the importance of their role in communicating policies and objectives to their staff and in communicating staff attitudes and reactions to their superiors. Good communication is especially important in an undertaking like telecommunications with large numbers of staff working in small and sometimes isolated groups.

Industrial relations

The approach to industrial relations depends very much on the practice and traditions of individual countries, and it would be of little value to describe BT experience. As a general matter, however, a willing staff is always to be preferred, and people cooperate when they feel they or their representatives are being consulted about what happens.

BPOT distinguished between matters where consultation with staff representatives might be advisable but managerial decision was final, such as staff deployment and technical and other procedures, and matters which were appropriate for negotiation with individuals or recognized unions, like pay, working conditions and so on.

Summary

To sum up:

— Staff and their management have a decisive effect on the success or failure of a telecommunications enterprise.
— Broad predictions of forward staff requirements by skill groups should be generated as part of the overall planning process (Chapter 5).
— More detailed predictions are needed for personnel planning purposes. There are well-established techniques for calculating requirements for engineering and operating staff. Determination of clerical staff requirements is more difficult, but techniques have been developed.
— It is imperative that there should be adequate numbers of managers and professional engineers. Arrangements to ensure this including management training deserve close attention by top management.
— Properly organized training is essential. The Maitland Report reviewed a number of important matters affecting training practices in developing countries.
— Planned measures for improvement in the productiveness of all staff groups are important in most enterprises.
— Care is needed to ensure that engineering productivness is not over-emphasized at the expense of quality of service and workmanship. Good workmanship at the time of installation not only ensures lasting good service but reduces future maintenance costs due to faults.
— BT experience has illuminated a number of matters to do with engineering workforce organization.
— Good internal communication in both directions between management and staff is particularly important in telecommunications.
— The approach to industrial relations depends on the circumstances and traditions of individual countries.

Part II The technology and its products

9 The technology in outline

The underlying technology of telecommunications and computing has become as complex as that of almost any sector of modern activity. But the principles on which telecommunications products work are quite simple and largely familiar. It does not require a great deal of effort to master them in sufficient depth to provide an informed basis for management and policy decisions.

This chapter is designed to provide a background for the rest of the book and to help non-technical people who may have to make such decisions. It is not designed for technical readers.

Computing technology

Computing techniques pervade modern telecommunications so thoroughly that it is easiest to start with a brief outline of computing technology.

The first problem in understanding any specialism is usually the vocabulary. All industries and professions have their own. These are often hard for lay people to understand. What makes many of them particularly difficult is that they use a large number of familiar expressions but with very special meanings. The vocabularies of telecommunications and computing are like this. The expression 'mouse' looks familiar, but there is no way a lay person can deduce what it means to the computing community without help.

It helps to get a general idea of the technical side if one thinks of telecommunications and computing equipment as made up of a series of functional building blocks. This also makes the vocabulary problem a lot easier. It is not difficult to find words which can be used with their everyday meanings to refer to these building blocks quite accurately.

Many of the building blocks are quite familiar. Telephones, dials, keyboards, electronic screens, printers, radio sets, and many others are part of everyday life. But some of the less obvious ones are buried inside computers or telecommunications networks. Few people but specialists see them or are conscious of their existence.

Luckily some of the most important of these buried building blocks can be understood by analogy with the relationship between a caller and a human telephone operator. A caller has first to 'input' to the operator the fact that he wants to make a call, and the number he wants. The operator uses her short-term 'memory' to remember the number the caller wants; and a 'store' of information (which may be in a reference file or her head) about how to connect the call. When she uses her brain and hand to set the call up using her cords and plugs she acts a as a 'processor' for the call.

In just the same way (Figure 9.1) computers have input devices which enable humans to feed them with instructions and information. These are usually keyboards, although computers which can 'hear' spoken instructions are now available. Computers have processors which correspond to the operator in the role she plays in the example above. Processors carry out the activities which the computer exists to perform — calculating, sorting, comparing and so on. They also control the behaviour of all the other parts of the computer. They do all this in accordance with programs of actions. These programs are sets of instructions which the computer works through in strict order. Present-day computers work on electric currents, and it is the flow of these through their circuits which does all the work. Electricity moves very fast, by human standards. A program of ten thousand actions may be completed in a tenth of a second. But a computer does still take time to work, just like a human.

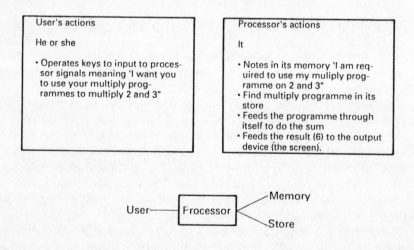

Figure 9.1 Simplified functioning of a computer

To help them, processors have short-term memories which are used to 'remember' temporary information such as programs in use and numbers to be multiplied. These memories also act as a sort of note-pad for the processors. For example, in the course of working out, say, '4 × 27' a human might say, 'Four sevens are twenty-eight; that is eight and carry two to tens; four twos are eight add two is ten; result 108'. 'Carry two' had to be held in his or her memory while the tens figure was calculated. Computer processors use computer memories in the same way.

Computers also have long-term stores which hold the programs and more enduring information for processing — lists of numbers, the text of documents and so on. Stores also hold the results of the processing. The technical name for such information is data. When such data (and other computing material like programs) are sent over telecommunications links, the process is called data transmission.

Finally computers need means to output their results — for example, '108' in the calculations above. This means is usually a screen. If a paper record is required there will also be a printer.

Digital and analogue principles

There are one or two other technical terms we shall need. The vast majority of modern computers and all the latest telecommunication systems work on the digital principle. Table 9.1 shows a very simplified example of a binary digital coding system. The ten letters a–j and the ten decimal numbers 0–9 are represented by a set of twenty eight-digit binary codes or bytes. The 1s and 0s in these codes are called binary digits. This way of representing information is extremely important. This is because the simplest way to vary electric current is to turn it on and off. If we make OFF represent binary 0, and ON represent binary 1, we can very easily put binary codes into electrical form. The circuits in a digital computer react to ON and OFF conditions, and therefore to information in the form of digital codes like this. The ONs and OFFs are called bits. It can be seen in Table 9.1 how each ordinary character has its unique binary translation.

It helps to think of a processor as an arrangement of electrically operated switches. Each switch can be turned on by applying electricity to it. If the electricity is turned off, the switch turns off. In a processor, large numbers of such switches (which are actually transistors or their equivalent in integrated circuits) are arranged in such a way that if, to take a very simple example, one programme of electrical ONs and OFFs is applied to them the processor will multiply; if another, it will divide.

Conventional telephones work on a different principle. The speech currents generated in the microphone in a telephone handset and the corresponding currents which are used to create audible speech in the receiver are unavoidably analogue — that is they directly reproduce (or are the analogue of) the corresponding sound waves in the speech or other noises being transmitted. The primary function of all older telephone networks and the plant in them was to carry these currents through to the distant end with a minimum of weakening, distortion and interference from noise. The techniques needed to do this are called the techniques of analogue transmission. Telephone exchanges designed to connect such currents are called analogue exchanges. At present in many countries the exchanges and inter-exchange links and the telephones themselves are still analogue, although as we shall see this is rapidly changing.

Table 9.1 A simple binary code

Letter or number	Binary code
0	00000000
1	00000001
2	00000010
3	00000011
4	00000100
5	00000101
6	00000110
7	00000111
8	00001000
9	00001001
A	00001011
B	00001111
C	00010000
D	00010001
E	00010011
F	00010111
G	00011111
H	00100000
I	00100001
J	00100011

The latest exchanges and links use digital techniques. The principle of digital telephone transmission is that the speech currents from the telephone are converted into digital codes and sent through the network in that form. This conversion can either occur in the telephone or other terminal or at the exchange. Exchanges which switch telecommunications traffic in this form are called digital exchanges.

We shall also need to use two technical expressions which describe the capacity of analogue and digital telecommunications channels. The capacity of an analogue channel is expressed by stating the frequency bandwidth — corresponding to the number of sound waves per second of the highest pure note it can carry. An ordinary telephone channel is designed to carry waves up to 3,400 per second. This is enough to make telephone conversation fully intelligible. One wave per second is called one Hertz. Thus, for example, a normal telephone channel is said to have a bandwidth of 3,400 Hertz or 3.4kHz. MHz represents millions of Herz, and gHz thousands of millions. The same measure is used to express the frequency and the bandwidth of radio transmissions.

The capacity of a digital channel is expressed by stating the number of bits per second (bit/s) it can carry. In present networks speech is usually sent at 64,000 bits per second or bit/s. This is normally written as 64 kilobits per second or kbit/s. Mbit/s means millions of bits per second and gbit/s means thousands of millions.

There are two other computing terms we shall need. Hardware means the physical equipment and components which make up computers, exchanges, terminals and so on. Software is the industry name for the programs of actions which determine the behaviour of this hardware, and are stored and used in the form of digital codes represented by electrical ONs and OFFs.

So much for terminology. We are now ready to use it to describe computing hardware and software, and telecommunications equipment.

Hardware

Electrical computing hardware has always had at its heart the active electronic devices — the valve, the transistor and the integrated circuit.

The earliest electronic computers used valves. But it was only when suitable transistors became available in volume production in the early 1960s that computing began to develop as a major industry. The invention of the integrated circuit (i/c) has proved to be a key advance. In a modern i/c there may be the equivalent of 100,000 or more transistors on a piece or 'chip' of silicon, about 5 mm square.

Techniques for forming more and more transistors and their associated circuits on a single chip advance continually. In very large-scale integration (VLSI) as many as a million or more transistors and the associated circuits are formed on a single chip.

Techniques for manipulating the physical and chemical structure of the silicon itself are also advancing remarkably. A number of different ways of permutating microscopic layers and patterns of silicon have been devised. There are real prospects of further advances using silicon devices operating at extremely low temperatures. At the same time rapid progress is being made with devices using other materials such as gallium arsenide. These devices make it possible to carry out millions of calculations per second. (Gallium arsenide is also very useful at milli-metric radio frequencies — see Chapter 11.)

There have been similarly dramatic advances in long-term storage. The long-term stores in the earliest computers consisted of magnetic tape or drums with a magnetic coating, working on the same principle as a tape recorder. Modern computers also frequently use magnetic stores. But in the 1950s a magnetic tape stored on a reel 30 cm in diameter and requiring a cabinet the size of a cupboard could store 800,000 bytes of information. At the time of writing, one magnetic disc in a box 10 cm square can store 40 million bytes. A larger disc can store 360 million bytes. New storage techniques providing ever greater user capacities, by using lasers, for example, continue to be devised and perfected. They will certainly lead to constant gains in performance and cost.

Mainframe computers

These developments have been exploited in some extremely significant advances in the size and physical characteristics of computers themselves. Originally, computers were large assemblies of equipment, usually housed in a series of cabinets and needing complete rooms to contain them. A typical computer in the late 1950s could process a few hundred instructions per second. It cost at the very least several hundred thousand pounds and occupied a whole room. Operators gained access to these machines by using separate terminals. These consisted of a keyboard, a screen and some simple electronics which could exchange signals with the computer. The terminals were normally located in a separate room.

In the 1960s and 1970s these big mainframe computers were used primarily on large-scale repetitive processes like preparing bills and accounts, scientific and engineering calculations or 'models' of the financial or other behaviour of businesses and activities such as transport systems. They often had a considerable number of terminals, provided so that details of, for example, amounts of money received and paid could be fed in by teams of specialized operators.

Minicomputers

In 1964 Digital Equipment Corporation in the United States introduced the first minicomputer. This was a machine of comparable capability to a mainframe but it exploited transistors to make it much smaller and cheaper. Minicomputers occupy cupboard-sized cabinets. But they are still specialized devices which usually need correspondingly specialized people to program them and set them up.

Microprocessors and microcomputers

The integrated circuit was invented in 1959. Very important advances were made in the United States in the 1960s in applying integrated circuits to computing. The key achievement was to create a complete processor on a single chip. The first commercial mircroprocessor of this kind ws introduced by Intel Corporation in the United States in 1971. Complete microcomputers incorporating microprocessors soon appeared. By 1975 the microcomputer was firmly established as a commercial proposition. A typical modern microcomputer has comparable processing power to the mainframe computers of the 1960s. But its processor and all the associated electronics fit into a box which may be less than 40 cm square.

Printers and other items of specialized hardware are frequently physically separate from the main computer or microcomputer. Collectively such separate devices are called peripherals.

Microcomputers are cheap enough and small enough to be used by individuals and to stand on desks. When used in this way they are called by the familiar name of personal computers (PCs). Another common use of microcomputers is as word-processors (WPs). These are specialized microcomputers whose computing power is used primarily to create and correct text before it is printed. Desk-top publishing is a technique which uses PCs to create finished documents with sophisticated type faces, diagrams and pictures comparable with normal printed material.

Before long the industry realized that it was possible to produce even simpler and smaller microcomputers, which would still have considerable power. One family of this kind are designed so that an ordinary television set can be used with them as a screen and a conventional tape recorder can be used as a store. Such devices are now familiar as home computers. A second family are genuinely portable lap-top microcomputers A typical lap-top computer measures about 25 cm × 15 cm and weighs only a kilogram or two. Yet it has processing, memory and storage capacities wholly sufficient for serious use.

Supercomputers

At the other end of the scale, the same advances in basic technology have been used greatly to improve the design and reduce the cost and size of mainframe computers. They have also been used to create a new family of supercomputers. These are machines of truly enormous power. A typical supercomputer can process millions of instructions per second. They are used mainly for scientific purposes or for specially demanding applications like weather-forecasting. But similar technology has made it possible to tackle some tasks hitherto beyond the scope even of computing. A good example is the storing and searching of fingerprint records. This involves a gigantic amount of information and processing because of the techniques that have to be used to record the detail of prints. Practical fingerprint systems are now on the market.

Artificial intelligence

An important recent development involves changing the way computers operate. Up to now, all computers have worked through a single set of instructions, step-by-step in sequence. The program might have 'loops' and other diversions, but the computer always followed them by 'putting one foot in front of the other'. Major effort has been devoted in recent years to techniques which would enable processors literally to do several things at once, as the human brain does. These techniques make possible computers with artificial intelligence, which can draw inferences like human beings.

Software

The term software was explained earlier. It refers to the programs which govern the behaviour of computers. We have already seen how numbers can be expressed in binary digital codes made up of ONs and OFFs. If, for example, codes of this kind representing '2' and '3' are fed into a processor whose switches are programmed to multiply them, the binary code representing '6' will appear at the output. Coded information intelligible to people, such as '2', '3' and '6' in the example is 'data' as we defined it at the beginning of this chapter. Programs and data can be stored in the computer's store, and temporarily in its memory. Programs, and therefore the processes they control, are capable of literally unlimited expansion and elaboration, subject only to the physical capacity of the processing and storage hardware available.

Only specialists can give instructions directly in the form needed to operate the processors in computers and microcomputers. To continue with an extremely simple example, a non-specialist user might by told to key 'multiply' when he wants the computer to do that. The signals sent by a typical keyboard to represent multiply would be eight-digit binary codes representing the letters M,U,L,T,I,P,L,Y. A processor needs a totally different pattern of ONs and OFFs to be set to multiply. Computers therefore contain a sort of dictionary to translate MULTIPLY into the corresponding processor control pattern.

A set of vocabulary intelligible to humans and which the computer can also understand is called a language. One could arrange that ordinary English (or any other normal language) be used to talk to computers. But it would be very complicated and wasteful to do this. Luckily it is unnecessary. Humans can readily understand simple stylized languages, provided they use ordinary letters and symbols. Computer languages are therefore very simplified and stylized. There are many different computer languages tailored to various fields of activity.

Programs and programming

The creation of software from scratch is a formidably complex and demanding task. Programming the large mainframe computers of the early years was a very specialized and laborious business. Vast amounts of effort were involved. The nature of the programming process was such that users' requirements had to be specified in great detail and with great precision. A very closely defined procedure therefore grew up, designed to hold down the time and cost of programming. The whole process took a long time. Even now it is quite common for programs for big accounts or billing computers or for the computer controls in modern telecommunications exchanges to take a year or more to perfect. A sector of the industry consisting of firms specializing exclusively in these problems grew up. They were called software houses.

Such procedures for creating tailor-made programs are acceptable for large-scale accounting, scientific or engineering processes involving enormous numbers of identical transactions, carried out in standard ways against stated requirements. But they discouraged use of computers for other purposes. It was possible to program computers in this way to keep accounts and file office records, for example. But it was a cumbersome and expensive business, and only big organizations found it worth contemplating.

We have seen how hardware developments from the late 1960s onwards created the potential for an immense expansion of the application of computing. The desire to exploit this potential, particularly the desire to open up markets for microcomputers, required a fundamental rethinking of the industry's approach to software. Standard programs were created for microcomputers which could be used as they stood by non-specialist staff, with only a minimum of instruction. Such programs could perform a wide range of office and similar processes — preparing profit and loss accounts, balance sheets, sales graphs and so on.

Software determines the nature of the dialogue between the user and the computer. Obviously the ease or difficulty of this dialogue greatly affects the value of computers and microcomputers to lay users. For a long time this dialogue took second place to the internal problems of computing technology proper. But in the last few years the industry has realized that the rigid and awkward dialogue has constrained the application of computing in many fields. Various ways have now been found to make this dialogue more human, or 'user friendly'.

Telecommunications plant and equipment

Finally we should look briefly at telecommunications equipment. Up to the late 1960s its design was regarded as a separate discipline from computer design.

Earlier techniques

All the original automatic telephone exchanges consisted of switches, selecting devices and so on which used electro-magnets to operate their moving parts. These electro-mechanical techniques were greatly elaborated and perfected over the years. In the late 1950s, for example, most advanced countries, including Britain, found ways to adapt electromechanical exchanges to enable callers to dial, first, their own trunk calls and, later, their own international calls.

In the earliest telephone networks, calls were carried or transmitted between exchanges over ordinary pairs of copper wires. Over distances of more than a few miles, this became a difficult and expensive business. If the speech signals were to reach the other end of the line with sufficient strength, the wires had to be very thick. When the valve was invented, it became possible to restore the strength of the currents by amplifying them, so that thinner wires could be used. In the 1920s and 1930s this made a marked difference to the cost of long-distance telephone calls.

Valves also made it possible to transmit a number of calls over a single pair of wires, by the technique called frequency division multiplexing. This works on the same principle as a radio set. A large number of different conversations are sent at different frequencies over one pair of wires. They are separated at the distant end by tuned circuits very like the tuning circuits in a radio.

The range of different frequencies which can be sent in this way over ordinary pairs of wires for any useful distance is limited. A much wider band of frequencies can be transmitted over coaxial cable, in which one wire is enclosed in an insulating sheath, itself in turn enclosed in a metal sheath. Point-to-point radio can be used in the same way as coaxial cable. It is possible to carry as many as 10,800 conversations on a single coaxial tube and 2,700 on a point-to-point radio channel. Analogue plant of both these kinds is still in very widespread use today.

Digital telecommunications techniques

All modern inter-exchange transmission equipment employs the technique called pulse code modulation (PCM). This uses the digital principle described earlier.

Digital techniques have very important advantages from the telecommunications point of view. A digital link has only to transmit digital code signals (in bits) with sufficient faithfulness to allow the values they convey to be read by the equipment at the far end. This is much easier to arrange than tolerable end-to-end transmission of the original analogue waveforms. Moreover, the analogue waveforms recreated at the far end of a digital link are, as it were, wholly new. They have not been weakened or distorted by transmission through the network and they are much freer from noise. PCM digital transmission links are cheaper and produce better quality signals at the far end than their analogue equivalents.

Digital telecommunications techniques have other very important advantages. Binary code signals of the kind they use are of course identical in form to the signals that circulate within and between computers. From the earliest days of electronic computing it was realized that if digital signals were fed into suitably arranged computing equipment, they could be routed to their destinations and otherwise processed in exactly the way required for a telephone exchange. Moreover if electronic exchanges of this kind were equipped with suitable software and stored instructions they could accept and interpret a much wider range of combinations of signals from telephone dials and keys than electro-mechanical exchanges. They could also communicate a wide range of control signals between their processors, and could handle changes in routing, to add new customer lines and so on very readily.

Because of the immense versatility of processors, these signals could be used to control a whole host of other facilities and services. Such electronic exchanges are called stored program control (SPC) digital exchanges.

Computing techniques and the associated active and storage devices have proved to be very well suited to all the processes that have to be carried out in an exchange. From the hardware point of view, modern telecommunications exchanges are very similar to mainframe computers. A typical digital exchange is much smaller physically than its analogue equivalent, and costs less.

Digital communication can be extended right out to the customer, over what is called the 'local [wire] loop'. At present the great majority of local loop circuits between customers and the exchanges that serve them still operate in the analogue mode. Data and other signals emerge from computers and computing terminals in digital form. Before they can be sent over these analogue circuits they have to be converted into analogue currents. They also have to be converted back at the other end. Devices that do this are called modems. Their design has been steadily improved over the years. But the process of conversion and back again is clumsy and expensive, and constitutes an additional source of failures and errors. If the local loop circuits and the rest of the network are digital, no analogue to digital conversion is required and these problems do not arise.

Optical fibres

Techniques for transmitting light pulses over very thin flexible glass (optical) fibres for telecommunications transmission have advanced extremely rapidly in recent years. Optical fibre cables are particularly suitable for carrying digital code signals. The highest capacity coaxial digital transmission links carry 565 Mbit/s. Links which carry 1.2, 2.4 or even 4 gbit/s are coming on to the market. A recent development has been devices capable of being coupled directly to three or more optical fibres, and of switching light signals between them in the same way as a telephone exchange switches speech.

Optical fibre cables have potential not only for inter-exchange links but also in distribution cable networks connecting users to exchanges.

Optical transmission and switching techniques are now being applied also in computing. Computers which work on light rather than conventional electricity have been shown to be practical. They can operate very fast indeed. They could well result in yet another major advance in the power and application of computing techniques.

This chapter has outlined the basis of the technology — hardware, software and telecommunications equipment. We can now turn to the application of this technology and to a number of policy and management issues to which it gives rise. We shall do this under six headings:

— the various kinds of telecommunications networks (Chapter 10);
— the role and exploitation of radio (Chapter 11);
— the planning and management of the plant that comprises these networks (Chapter 12);
— international plant and services (Chapter 13);
— the apparatus on customer premises (Chapter 14);
— services which add value to the use the customer gets from the network (Chapter 15).

10 Networks

Each country's telecommunications networks form part of the combined world network, which extends to most of the inhabited surface of the globe. It is the largest human artefact. It links over 150 countries and serves over 600 million customers.

Public Switched Networks (PSTNs)

In all countries the largest and most important network is the Public Switched Telephone Network (PSTN). Many PSTNs are made up of old and largely worn-out plant. For example, in March 1988, 2,758 of BT's 6,866 local exchanges were still Strowger electro-mechanical exchanges. Many such older exchanges give indifferent service. Calls fail and when connections are made they are often noisy. Computers are particularly intolerant of clicks and bangs. But because the PSTN is universal it is used to carry a great deal of data traffic. Technology to correct the resulting errors has been carried a long way. But in most countries the maximum digital bit rate that can be sent over the analogue PSTN is 9.6 kbit/s and even that does not always work properly.

Like an increasing number of advanced countries and many developing countries, BT has a major programme to modernize its PSTN, using SPC digital exchanges and digital inter-exchange links. In 1989 this programme is well on the way to completion.

Telex network

The telex network is usually separate from the PSTN. It has its own exchanges and its own circuits between them. Telex circuits may pass physically through transmission systems, cables and other plant provided primarily for the PSTN. But they are electrically quite separate. In common with many advanced administrations BT is in course of replacing its electro-mechanical telex exchanges with SPC exchanges. It is important to secure a return on this modernization and BT is vigorously promoting telex. The BT telex growth rate has recovered notably in the early 1980s.

Nevertheless by modern standards telex has important limitations. Only upper- or lower-case letters can be sent at any one time. Conventional telex operates at only 50 bit/s. Compared with the speeds of computers and PCs this is extremely slow. Nevertheless in Britain at any rate the telex network is quite extensively used for sending data between computers.

Data transmission: packet-switching

Although PSTNs and telex networks are extensively used for data transmission neither of them is really suitable for this purpose. The simplest dedicated network for data traffic is one with high-quality, 'quiet', error-free exchanges and circuits capable of really high bit rates, in which through circuits are set up just as in the PSTN. This kind of network, which is called 'circuit-switched', is common in advanced countries. West Germany, for example, has created several circuit-switched networks for varying speeds of transmission. But in Britain, North America, France and other countries advantages were seen in another approach, called 'packet-switching'.

For transmission over a packet-switched network data are broken down into a series of blocks or packets. These are of standard length and format. Each one includes the address of the destination as well as part of the data. Packets are typically 1,024 bits long and each one takes only a very short time to send. Each packet is handled independently of the others in the message. Processors in the network tell the source computer every time they are ready to accept a packet. Once it is released they read the address, decide the path it is to take and send it on its way. Packet techniques make it possible very much to reduce errors in transmission. They have now been perfected to the point where satisfactory speech can be sent over packet networks, but they remain primarily vehicles for data transmission.

Because each packet waits until a link through the network is free, the control processors are able to pack the links with traffic. If there is a breakdown messages can be sent over routes which avoid it. Terminals operating at different speeds can communicate over a packet network. The main features of such a network are illustrated in Figure 10.1. For certain kinds of data communications packet-switching has real advantages over circuit-switching. These advantages increase markedly with distance. Over international distances they become really important.

Key

E1 =	Equipment for assembling packets and controlling their release	N =	Part of message being transmitted
E2 =	Equipment for receiving packets and resembling original message	P = S = 1–5 =	Part already transmitted Packet switches Packets corresponding to N
M =	Untransmitted part of original message awaiting transmitted		

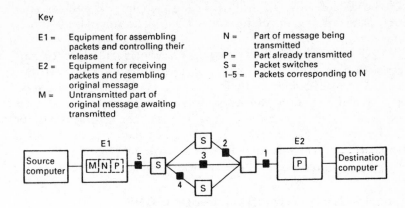

Figure 10.1 Simplified packet-switching network

Integrated Digital Networks (IDNs)

We should return now to the PSTN and its modernization. It is possible to introduce digital plant into an analogue PSTN in a piecemeal way, with calls passing over a mixture of analogue and digital plant. The real value of digital systems in terms of cost, quality of transmission and freedom from noise cannot be realized, however, until all the plant in the path of a call from the entry to the local exchange at one end to the point of leaving the local exchange at the far end is digital. Once encoded, the traffic can then stay in digital form till it gets to the other end; and the various processes that occur inside the digital installations can be synchronized. Such a network is said to be an Integrated Digital Network or IDN.

Once an IDN has been created, it provides the opportunity for another kind of integration. We have seen how speech can be converted into digital signals. Computer signals are already digital. Signals from teleprinters are digital, though they need special processing to pass over a modern digital network. Facsimile terminals and others can also be arranged to transmit digital signals. The great majority of forms of telecommunications traffic could therefore go over a single digital network.

Generally speaking, the more telecommunications traffic is concentrated the cheaper it is to handle. A unified network for all kinds of traffic is therefore an attractive idea. The evolution of the IDN is illustrated in Diagrams 1–3 in Appendix 5.

Most advanced telecommunications operators have already decided to create IDNs as their main communications vehicles for the future. But there has to be a process of evolution. No operator, however wealthy, could possibly afford wholesale replacement of all his plant overnight. Even if he could the practical, manpower and other logistic problems

would be insoluble. But the economics are such that it is now possible to contemplate replacement of the strategic parts of a national network over a few years.

We have seen that an IDN can carry any kind of circuit-switched traffic. Given the right software it could also function as a packet network. But in many countries the telex and packet networks are either modern or in course of modernization. The investment they represent has to earn a due return, and in many cases it is likely to be some time before their traffic is passed to the IDN. But the IDN will have to be geographically universal if it is to carry speech. It seems likely that it will increasingly become the principal communications vehicle for all kinds of traffic as the years go by.

Integrated Services Digital Network (ISDN)

The IDN as described above starts and finishes at the local exchange. Traditional customer telephones and the circuits in local cables that link them to their exchanges are analogue. If a computer or a PC is used on an ordinary telephone line a modem must therefore be used to convert the digital signals into analogue form. It was a natural step to develop technology which would carry digital service out to the customer's premises. Modems would no longer be needed for computers; although telephones would have to include circuitry to convert analogue speech waveforms into digital code. All services would then be digital throughout and use the same circuits back to the exchange. This arrangement is called the Integrated Services Digital Network or ISDN.

The expression ISDN is used primarily to describe what the customer 'sees' and the way traffic is routed between him and the exchange. The way traffic is routed beyond the local exchange is strictly speaking a separate matter, determined by the layout of the inter-exchange network on the ground. Telephone and circuit-switched data calls may be passed forward over what is in essence a speech network; telex calls may go over a separate telex network; and packet-data calls over a separate packet network; or some or all of these categories may pass over a combined IDN. But if all traffic leaves a customer's premises on a single group of digital circuits he is using the ISDN.

The techniques used to extend digital transmission into customers' premises depend on the size of customer installation which is involved. PABXs (see Chapter 14) and larger complexes can be linked to their local exchanges using 2 Mbit/s, 30-channel digital systems closely similar to those used on inter-exchange links inside the public network. Such systems provide signalling between the PABX control processor and the local exchange processor over a common channel independent of the speech channels. Specialized signalling systems have been developed for this purpose — for example, DASS2 in the United Kingdom. Such 2 mbit/s links may be provided over special metallic pair cables, over coaxial cables or over optical fibre cables.

A different technique has been developed in connection with ISDN service to customers with more limited communications needs. A pair of wires in an ordinary local telephone cable in good condition can be arranged to provide a bit rate capability of 144 kbit/s to customers up to about 5 km from the exchange. Of this, speech needs only 64 kbit/s. Another 64 kbit/s can be used simultaneously for a completely separate service — say, signals from a PC — while still leaving a 16 kbit/s channel available either for yet another service — say, another computer or for control signals from the terminals to the exchange. The technical name for this technique is 2B+D. Two people holding a conversation over a 2B+D link could therefore exchange material between their personal computers or facsimile machines over the same line at the same time. Documents or diagrams under discussion could be exchanged and altered simultaneously at both ends during the conversation. Once the necessary equipment has been installed at the exchange and at the customer's premises the 2B+D technique could enable two extra channels — 64 and 16 kbit/s — to be made available over an ordinary telephone line without the need to lay any new cable at all.

This is potentially an important development. But the transmission requirements for high bit rate digital signals are exacting. Much depends on the character and condition of individual cables. Old cables, cables in poor condition or containing too many permanently useless and disconnected pairs may not be able to carry 144 kbit/s. This is fundamental to the whole idea. And there is a limit to the number of customers who can be served at 144 kbit/s over the same cable.

Two other recent developments concerning digital service to customers should be noted. Where optical fibre or coaxial cables are or can be made available to customers' premises they can be linked to the exchange over circuits operating at higher bit rates than 64 kbit/s. In principle circuits can be provided operating at any standard bit rate up to the limits of the technology. Speeds of 256 or 512 kbit/s, 2, 8, 34, 140 (for full broadcast quality television) or 565 mbit/s, or 1.2, 2.4 or higher Gbit/s are all available. Switches are being developed capable of handling these very high bit rates. From the technical point of view the possibility already exists of creating a Broadband version of the ISDN. This concept has been developed in West Germany, among other advanced countries.

The second development concerns the ability to transmit speech at bit rates less than 64 kbit/s. Acceptable techniques are now available to do this. Bit rates of 32 kbit/s and even 16 kbit/s are used in specialized applications like the transmission of speech over packet networks and, significantly, in mobile radio systems. Several manufacturers have vision-phones (Chapter 15) in development which send speech at 16 kbit/s and pictures at 48 kbit/s over the same 64 kbit/s channel. The 48 kbit/s facility could of course be used for data, facsimile, text or any other traffic. Potentially, very similar facilities to those offered by 2B+D could be provided over a single 64 kbit/s switched channel. These techniques have

Photograph 1   Changing the constitutional status and internal structure of a PTT is a big task, calling for full time attention by senior officials (see page 11). Mr M O Tinniswood, Director, Reorganization Department, General Post Office.

Photograph 2   BPOT enlisted the help of distinguished accountants from business to help recast its accounting systems (see page 33). Mr W P Kember, in the later 1970s Senior Director Telecommunications Finance BPOT and now a senior finance official in BT plc, had worked previously in Coopers and Lybrand and British Oxygen.

Photograph 3 Any forecast is liable to be wrong and mobile equipment can be of great value in meeting pressing shortages of capacity (see page 129). A British digital (System X) mobile exchange unit is lowered into place.

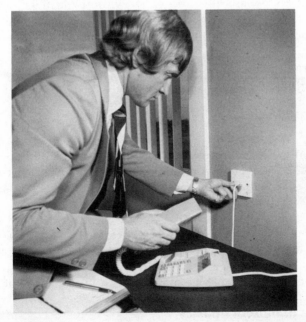

Photograph 4 The introduction of plugs and sockets radically advanced telecommunications customer premise equipment in the UK from the points of view both of design and of commercial arrangements (see page 143).

still to be perfected, but it is clearly possible that they could supersede 2B+D.

The commercial prospects for all these developments involving two or more bit streams over one pair of wires are, however, far from clear even in the advanced countries. 2B+D facilities are offered by most major manufacturers on PABXs, where they are relatively easy to provide; and PABXs are increasingly linked into public networks by 2 mbit/s systems. But public 2B+D services are only just beginning to be introduced and it is not clear what the demand for them will prove to be. 16+48 kbit/s systems are still at the development stage. The market for Broadband ISDN is equally uncertain (Cable Television systems are dealt with separately below).

A key question concerns the cost and complexity of customer apparatus for digital customer service of any kind. The advantages of the IDN configuration are generally recognized. But it is still not clear that the advantages of digital service out to the customer, in the shape of the elimination of modems and of facilities like 2B+D outweigh the extra cost of digital rather than analogue operation on the customer link. The heart of the problem is the cost of integrated circuits in the telephone or other terminal to convert analogue speech waveforms into digital code and vice versa; and the fact that the exchange line termination for digital service costs about the same as that for analogue service.

Most developing countries would probably do best to concentrate on creating IDNs, where these can be shown to be justified, and to leave digital customer service and more complex applications like 2B+D and Broadband ISDN until the economics and demand for the facilities are clearer.

The various stages in the evolution of the ISDN are set out in diagram form in Appendix 5.

WANs and MANs

The computing industry uses the term Wide Area Networks (WANs) to refer to inter-computer networks established over significant distances. Some WANs extend between continents. They often use telecommunications leased circuits as transmission media; or they may be set up using the PSTN or public packet-switched services; or any permutation of these.

The expression Metropolitan Area Networks (MANs) is used to refer to broadband networks installed primarily to interconnect computers in big cities. MANs may also be used to carry television signals and speech. A MAN of course has much in common with telecommunications plant.

Changes in network basics

The changes in the technology are leading to important changes in some of the basics of telecommunications networks. Conventional analogue networks are laid out in strict patterns. These are designed to meet the requirements of analogue transmission technology and to optimize the way plant is laid out and calls are routed from the point of view of operating economics. The maximum loss which would be acceptable if callers were to be able to hear one another properly was determined by experiment some years ago. Rules were developed for the loss in strength of signal on each link of an analogue network which would be appropriate from the technical point of view. From these definitions and rules administrations laid down an optimum structure for the layout of their exchanges and links and for transmission losses and the routing of calls. This structure had to be strictly adhered to if the network was to work properly. Such structures are universal in countries with analogue networks.

In wholly digital networks with processor control such restrictive rules for routing are no longer necessary. Transmission losses on end-to-end connections are for practical purposes eliminated and calls can take many different paths through the network.

Processor control creates flexibility in many different ways. The concept of distinct local, trunk and international exchanges is disappearing. In a digital network international calls, for example, may be switched at points in the network anywhere from a few hundred metres from the customer's premises to outgoing international terminals like satellite earth stations. And these switching nodes are likely themselves to be controlled from remote centres. Finally, the traditional switching function is increasingly likely to be taken over by 'intelligent multiplexers' which can group and re-group circuits as required under processor control. The role of traditional local exchanges is likely to pass to small 'intelligent' units integrated into the customer distribution network.

The 'intelligent' network

Digital networks can be actively managed from minute to minute. When things go wrong the control processors can steer round failures or indeed contain and even cure the problem. They can change the routing of calls as the load on the network changes. Alternatively traffic patterns and events like breakdowns and congestion within the network can be monitored by computers and brought under control by human intervention. Everything that happens throughout the country on a modern network down to and including provision of and events on customer distribution circuits will be monitored and controlled from a single computer-based Network Control Centre. New circuits and traffic routes will be created and modified by remote control.

Central processors and databases within the network, with appropriate signalling facilities back to traffic exchanges, can provide a much wider range of services than just call connection. For example, for several years

BT has used electro-mechanical equipment to provide the '800' automatic freephone service. A customer who dials 800 plus an appropriate selecting code is connected to a distant line free of charge. The cost of the call is charged to the recipient. This service, with variants under which the charge is shared between originator and recipient, is very widely used by advertisers and others and its use is growing.

Similar facilities can be provided more quickly and cheaply by central computing equipment. Such equipment can also provide true 'transaction' services like automatic credit card verification. A network which has such capabilities is called an 'intelligent' network.

The Universal Intelligent Communications Network (UICN)

At the present time in the light of developments like these, historical concepts of telecommunications network structure and economics are being completely recast. Eventually the network will probably be completely controlled by artificial intelligence. The Universal Intelligent Communications Network (UICN) will succeed the ISDN.

Cable television techniques

In the last few years the techniques used in local cable television (CATV) distribution have been evolving along parallel lines to those used in local telephone cable networks. CATV networks have existed to carry programmes otherwise available off-air in a number of countries including the United States for many years. They have not been widespread in Britain, probably because of the high quality of the direct broadcast programmes which are available.

It was realized that such networks could be upgraded to have the capacity to carry additional programmes. This led to important developments in the technology. In their original form they were simple one-way systems which delivered all the programmes on offer to selector boxes on top of customer's sets. These are called 'tree-and-branch' systems. A large number of such systems are in service in advanced countries.

It was felt, however, that there was a good case for putting switches further back in the network. The original cables could carry only, say, half-a-dozen programmes. To deliver thirty programmes to each house required more expensive cable. If switches could be located at intermediate points in the network and controlled by customers, the cables into each customer's home need only be cheap ones capable of carrying from one to, say, four programmes (depending on how many sets were to be served). Such systems are called 'switched-star' systems.

Alternatively, broadband 'tree-and-branch' optical fibre or coaxial cable systems laid right into the home are now used to deliver thirty or more

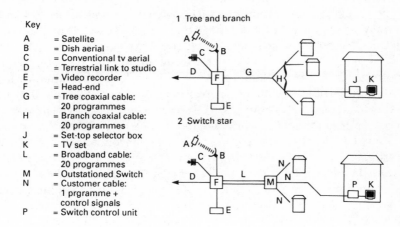

Key

A	= Satellite
B	= Dish aerial
C	= Conventional tv aerial
D	= Terrestrial link to studio
E	= Video recorder
F	= Head-end
G	= Tree coaxial cable: 20 programmes
H	= Branch coaxial cable: 20 programmes
J	= Set-top selector box
K	= TV set
L	= Broadband cable: 20 programmes
M	= Outstationed Switch
N	= Customer cable: 1 prgramme + control signals
P	= Switch control unit

1 Tree and branch

2 Switch star

Figure 10.2 Entertainment networks

programmes without switching except at the set. The basic principles of the two kinds of cable television networks are illustrated in Figure 10.2.

The development of these techniques to deliver television to the customer is likely to be influenced in basic ways by the development of High Definition Television (HDTV). This is a system in which television cameras and receivers operate at much higher definition standards — typically over 1,000 lines compared with 525, 625 or 819 lines in existing systems. The objective is to deliver a much higher definition picture to the customer's screen, which in turn can be made larger. The 'aspect ratio' — the ratio of picture width to height — can also be improved. HDTV signals are likely to require higher bandwidth or bit rate channels than present television.

The idea that CATV customers might have terminals which could send signals back into the network led to another step in thinking. If signals could be sent to select programmes, they could be sent for other purposes. Speech, computer signals and so on could travel over such networks in both directions. The idea crystallized of switched broadband networks, capable of carrying all forms of telecommunications in both directions as well as TV programmes in the outward direction.

However, the technical requirements for such a network are very exacting. It has to meet the full needs of, say, thirty or forty entertainment channels in the outward direction and at the same time the different but equally exacting requirements of telecommunications in both directions. All this is possible with the technology, but it is very expensive. As a result there are few such networks as yet even in advanced countries. A number of interests are promoting such systems in Britain at present and they may become significant, although opinions differ widely on this. But it is doubtful whether any developing country except a very rich one could support such a network over any significant part of its territory.

More generally, penetration of CATV networks of any kind varies widely between countries. They are well developed and still expanding in the United States and in countries in Continental Europe like Belgium and the Federal Republic of Germany. But in Britain they are limited in extent and growing only slowly, apparently because the economics are still very uncertain.

Direct Broadcasting by Satellite (DBS) (see Chapter 11) is becoming an important competitor of orthodox CATV. Systems which combine a DBS receiver with a cable distribution system, say, to a housing estate or a hotel are appearing (the technique is called Satellite Master Antenna Television — SMATV). Terrestrial multi-channel microwave radio systems (MVDS) are being exploited as a cheaper alternative to cable-borne CATV. They operate at frequencies similar to or higher than those used by DBS and involve similar customer receiving apparatus.

Developing countries will make their own judgements about the case for CATV, SMATV or DBS on their territory. The important point in the present context is that in Britain combined CATV and telecommunications networks have yet to prove in commercially, even in affluent environments; although as we have noted there is considerable interest in their development at the time of writing (see also page 92).

Private networks

So far we have concentrated on public telecommunications and television services and the networks that provide them. Nowadays customers rent or own a wide variety of networks of their own — either wholly on one set of premises or spread over several different premises. The technology and layout of these networks have much in common with that of public networks. Private networks usually include one or more Private Automatic Branch Exchanges (PABXs) (see Chapter 14).

The conventional arrangement is for the cables and transmission equipment used to provide private circuits to be physically part of the plant provided primarily for the PSTN. Like the telex network, however, these circuits and the networks built up from them have been electrically separate from the public telephone and telex networks and from one another. There have been exceptions in a few instances where customers rent circuits for only part of the time, and the plant concerned is put back in the general pool for the rest of the time. But in analogue networks such arrangements are clumsy and not common.

Private customers are increasingly acquiring their own plant. Many private networks in advanced countries, for example, use microwave systems owned by the customers themselves.

In processor-controlled digital networks, however, circuits can be allocated between different applications from minute to minute, if this is required. In the IDN, therefore, it could well be common for end-to-end private circuits to be set up by processor-controlled switching as and when they are needed. Within the inter-exchange network permanently wired private circuits may become a thing of the past before long for this reason. Digital customer circuit techniques will equally make it possible to provide the equivalent of private circuit facilities in distribution plant without earmarking or wiring plant or channels permanently.

The effect of these developments is that nowadays 'virtual' private networks can be created, configured and changed entirely under processor control. WANs can be created in this way. The traditional distinctions between switched and non-switched (private) circuits are being steadily blurred and eroded.

Some idea of the scale and importance of modern private networks can be gained from the United Kingdom. In Britain it has been estimated that the 300 largest users have about 650 private networks between them, with one million telephones and 300,000 other terminals. In many cases, these British private networks form part of international — often intercontinental — private networks belonging to the parent companies concerned. An increasing proportion of the circuits on these networks are digital. They do not pass through public exchange switching plant. They are therefore generally much more suitable for computer traffic than the analogue PSTN. The 2B+D technique discussed above can be used on lines from digital PABXs to their outstations, as well as on lines to public exchanges. Digital facilities including 2B + D or 16 + 48 Kbit/s visionphones and other advanced applications are therefore being introduced within organizations and on private networks before the public exchanges concerned are digital.

These larger customers often have private telex circuits and networks in addition to their speech networks. Such networks frequently incorporate message switches. The nature of message and document transmission services which do not involve human conversation makes it possible to hold complete messages at central points in a network until the circuits beyond are free to accept them. Such arrangements have particular value for companies with extensive international message traffic. They make it possible for them to rent, say, a single circuit London–Tokyo, which can be intensively loaded with traffic which accumulates on the message switch until the circuit is free.

Summary

To sum up:

— In all countries the Public Switched Telephone Network (PSTN) is the largest and most important telecommunications network at present.
— In most countries the PSTN is still composed of analogue plant of varying age and condition. It usually gives indifferent service on telephone calls and it is far from satisfactory for data transmission.
— Telex is still expanding, but it is obsolescent.
— Many advanced administrations have introduced dedicated data networks. The British data network uses packet-switching.
— Most advanced countries are introducing Integrated Digital Networks (IDNs) to take over the role of their analogue exchange and inter-exchange networks. A suitable IDN can carry all kinds of telecommunications including text, data and images as well as voice calls over homogeneous 64 kbit/s channels.
— The Integrated Services Digital Network (ISDN) carries the IDN concept a stage further. Each ISDN customer will be served by a single group of circuits carrying all his telecommunications.
— Smaller ISDN customers and PABX extensions will be able to be served by the 2B+D technique, providing two 64 kbit/s channels and one 16 kbit/s channel over a single pair of wires.
— As an alternative to 2B+D, techniques are now available to transmit satisfactory speech at 16 kbit/s, leaving the rest of a 64 kbit/s channel available for other uses like data and moving pictures.
— The ISDN concept has now been developed into a Broadband ISDN, offering high bit rate switched channels as well as 64 kbit/s channels.
— The majority of developing countries who wish to develop their telecommunications would probably do best to concentrate on creating IDNs, leaving the full ISDN until its economics and demand are clearer.
— Cable Television (CATV) distribution techniques have converged with those of telecommunications. But the commercial viability of combined CATV and telecommunications distribution networks has still to be demonstrated.
— Millimetric Video Distribution Systems (MVDS) may offer a cheaper way of providing facilities equivalent to those of CATV proper.
— Satellite Broadcasting (DBS) and Satellite Master Antenna Television (SMATV) are becoming important competitors to cable distribution of television.

— It is not yet clear which, if any, of these technical approaches will predominate or become the technical vehicle for High Definition Television (HDTV).

— Private networks are becoming ever more extensive and elaborate. Many of them are international.

11 The role of radio

Radio techniques are familiar in many fields. They are the basis of the broadcasting industry. They are an essential element of military, aeronautical and marine activities. They have been exploited in telecommunications for many years and in many ways. A new impetus has developed in the use of radio in telecommunications in recent years. This chapter examines radio in telecommunications with particular reference to the needs of developing countries.

Spectrum problems

All radio communications use frequencies in the electro-magnetic spectrum. The range of usable frequencies has been greatly extended by technical advances. But it is still finite.

The principal sectors of the spectrum which are regarded as usable at present are:

— low frequency (long wave) 30–300 kHz;
— medium frequency (medium wave) 300 kHz–3 mHz;
— high frequency (short wave) 3 mHz–30 mHz;
— very high frequency (VHF) 30 mHz–300 mHz;
— ultra high frequency (UHF) 300 mHz–3 gHz;
— super high frequency (SHF) 3gHz–30 gHz;
— extremely high frequency (EHF) 30 gHz–300 gHz.

Frequencies between 300 mHz and 30 gHz are often referred to as microwave frequencies. Frequencies above 30 gHz are referred to as millimetric frequencies. Telecommunications have to share this finite spectrum with a great many other users of equal importance. Two transmitters on the same frequency interfere unless they are sufficiently far apart. The distance needed to separate transmitters varies with the frequency they use, the design of their aerials and the power with which they operate. Powerful long-wave broadcast transmitters must be separated by 2,000 km or more if they are not to interfere. Ultra-high frequency transmitters of the kind used in modern mobile radio systems operate at 900 mHz. They need to be separated by tens of kilometres if they are not to interfere. At any frequency there is a definite limit to the number of transmitters that can operate in any given tract of country.

Again, transmitters on different frequencies must use channels which are sufficiently separated in the spectrum. The extent of separation required varies with the kind of signal. Medium-wave transmitters in Europe are normally separated by 18 kHz. Mobile radio transmitters are separated by 12.5 or 25 kHz; VHF frequency modulation transmitters are separated by 200 kHz and so on.

At world level, the use and apportionment of the frequency spectrum is regulated by the IFRB (see Chapter 13).

Short-wave frequencies

The earliest uses of radio for public telecommunications were over long distances. The properties of the short-wave frequencies between 3 and 30 mHz are such that they can be used for communication over very long distances. These frequencies bounce off the ionized layers in space round the Earth, and in this way can be used for communications to any point on the globe. Until the first long-haul submarine telephone cables were laid in the 1950s, short-wave or high-frequency radio was the only way of providing transoceanic telephone service.

Short-wave techniques have become very sophisticated. But short-wave communications are rarely wholly satisfactory. The frequencies usable vary with the time of day. Short-wave tranmissions are very vulnerable to fading, distortion and interference. For international telecommunications HF systems are now largely superseded by submarine cables and satellites.

Microwave frequencies

Microwave frequencies between 30 and 300 mHz do not bounce off the ionized layers in space. They are generally only suitable for use over much shorter distances. For all frequencies above 30 mHz the range of transmitters of a given power decreases as frequency rises. The higher the frequency, the more its behaviour is like that of light. Frequencies above 300 mHz can be focused like light waves, and the transmitting and receiving aerials need to be visible from one another.

Where a short-wave link can carry only one conversation, a microwave system can carry thousands. Such a system has a similar communications capacity to the coaxial cables described in Chapter 9. Digital microwave systems can carry up to 140 mbit/s, and systems of greater capacity are being developed.

Microwave systems usually involve less extensive construction work than cable systems, although a large microwave station can cost millions of dollars. They have their own problems — for example, with the environment, because microwave aerials often need to be sited in prominent places. But they are usually quicker to install than cable systems and are much better suited to difficult country like mountains or jungle. Over a certain distance they become progressively cheaper than coaxial cable systems. For a given communications capacity, they have less cost advantage over optical-fibre systems. Subject to spectrum being available they are often well suited to most of the main transmission requirements of developing countries. But the choice between cable and microwave should be based on thorough economic studies of the individual case.

Satellite communications techniques

Satellite communications technology has made considerable advances in recent years. The original communications satellites carried low-power transmitters and receivers. They required large earth stations, with very big dish aerials about 30 m in diameter.

It is now possible to design satellites with more powerful transmitters and receivers. They can communicate on the ground with quite small dish aerials. These aerials and the associated electronic equipment are very much smaller and cheaper than those used with conventional satellites. Dishes only one or two metres across are used, for example, for one-way broadcast data reception. Small dishes for reception only — for example, for use in conjunction with Direct Broadcasting by (high power) Satellite — are even smaller and cheaper. Intense effort is being devoted to reducing costs still further. Progressively cheaper designs are coming on the market.

Satellite radio links are now used extensively by telecommunications operators throughout the world. They are particularly useful and cost-competitive over very long distances. About half the communications traffic across the North Atlantic is carried by satellite, for example.

In developing countries satellite communications have an important application in giving service to remote areas with a low density of lines and traffic. They are also used by countries like Indonesia to provide links to island and other locations which it would be much more expensive to reach with terrestrial links. The application of satellites in developing countries and remote areas is fully discussed in the literature (see GAS 8, 1983).

Satellites are increasingly used also for domestic telecommunications in advanced countries with large territories like the United States, Canada and Australia. For regulatory reasons there has been a special emphasis in the United States on satellite systems operated by carriers other than AT&T to carry services other than speech, especially data communications.

Millimetric frequencies

The frequencies above 30 gHz have their own disadvantages which are considerable. They behave much like light. Aerials must have line-of-sight one to another. The frequencies are more vulnerable to weather conditions, and there are certain points in the spectrum where they are specially liable to be absorbed by such elements in the atmosphere as oxygen. With present techniques aerials have to be located with great care, if dead spots and undue fading or distortion are to be avoided. But a lot of attention is being given to overcoming these disadvantages in advanced countries because there are some important advantages above 30 gHz.

In particular, the strength of signals drops off quite rapidly with distance. Transmitters can be located much closer to one another without risking interference. It should be possible to locate transmitters on the same frequency within a few kilometres of one another. Again, the bandwidth required for telecommunications and even television transmission is relatively small set against the capacity of the millimetric spectrum. A single gHz of spectrum can provide twenty-five colour television systems. In theory a large communications capacity is therefore available from bands already allocated in principle for telecommunications purposes above 30 gHz. Also, aerials are physically small at these frequencies, and less obtrusive from the point of view of planning and amenity.

Mobile communications

There have been important developments in mobile and portable radio communications in recent years. There are three main kinds of mobile communications systems for civilian road vehicles: mobile radio networks linked to the public telephone network, cellular mobile radio systems and private systems.

Private mobile radio networks have been operated for many years by emergency authorities such as police, fire and ambulance; by public utilities such as gas and telecommunications; by various commercial enterprises such as taxi services; and by vehicle fleet operators of many kinds. These private networks are exclusively for the use of their operators, staff and other participants — taxi drivers, for instance. They are not open for use by members of the general public; and they are not normally interconnected with the public telephone network.

Public mobile services linked to the public telephone network

In Britain BT introduced the first public mobile radio system linked into the main public telephone network in 1959. This service still operates. In its latest form, System 4, user vehicles are equipped with fully automatic terminals which behave very like normal modern telephones. Numbers can be keyed, and additional facilities, for instance, repeating the last number dialled, are provided. In Britain as elsewhere, however, the frequencies available for such public mobile services are very limited. Demand for BT's System 4 long ago exceeded supply, simply because, despite technical advances, there is a limit to the number of user vehicles which can be accommodated in the available spectrum.

Cellular mobile radio systems

There have been similar pressures for many years on public mobile radio in advanced countries generally. As a result, AT&T devised the cellular system. The first operational cellular mobile radio system was introduced in Chicago.

The principle of cellular radio is illustrated in Figure 11.1. The service area is divided up into a series of cells, which may be anywhere from one to about 15 km across. The fixed base station transmitters are located and arranged in such a way that no two adjoining cells are served by the same frequencies.

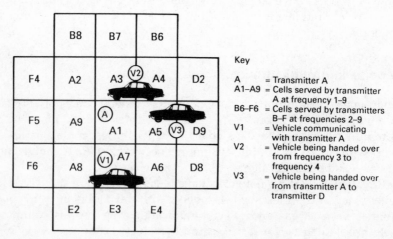

Figure 11.1 Principles of cellular radio (real cells are irregular in shape)

The operation of cellular radio depends on the fact that as vehicles move about, passing from cell to cell, they are handed over between base stations and frequencies. For this to be done, techniques had to be developed to detect the moment when a mobile crossed a cell boundary, to identify the frequency it needed to use to communicate in the new cell and to tune the mobile set to it. All this has to be done instantaneously, so that there is no perceptible interruption in a call if one is under way. Microprocessors with the appropriate software are used to do this. The resulting speech quality and general performance is much better than that of older mobile systems. Users are completely unaware that their calls are being handed over.

The power of the transmitters and the design of the aerials are such that the frequency used for Cell A2 in the example in Figure 11.1 can be reused for Cell B2, which may be only a few cell diameters away. The corresponding distance for conventional mobile systems like System 4 could be as much as 150 km. The result is that cellular techniques allow a very much greater number of mobile sets to be served with a given number of frequencies in any given tract of territory than conventional techniques.

The two UK public cellular radio operators, Cellnet and Racal Vodaphone Ltd, opened services in London early in 1985. Mobile sets are obtained from companies separate from the cellular operators.

Cellular radio has many potential applications besides straightforward mobile telephone service. Users can connect terminals such as portable PCs to cellular mobile sets and send non-voice traffic out into the cellular and main public telecommunications networks. Transportable terminals are available for use on both British services. These are self-contained mobile stations about the size of a small briefcase which can be carried about. They offer the same facilities and coverage as mobile sets. They are of great value on construction and other outdoor sites and in similar less permanent locations. Genuine handheld sets are also now available.

Cellular radio techniques have very important potential for developing countries. There is no reason in principle, apart from cost, why present cellular radio techniques should not be used to give normal fixed telecommunications service in areas where no cables exist, provided spectrum is available. The pressures on spectrum are normally less in less developed areas and there should therefore be plenty of frequency available.

Cordless customer premise equipment

Cordless telephones are an important development. They have become widely popular in Britain. They have base units corresponding to the main body of a conventional telephone, but without a dial or keypad or a handset. These base stations are connected to the main telephone network by wires in the normal way. The handsets incorporate keys. They are of much the same size and weight as conventional telephone

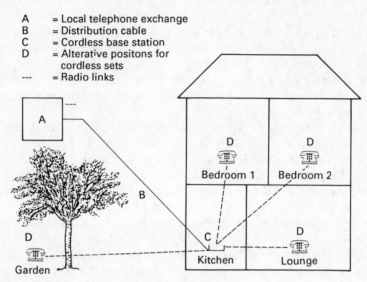

A   = Local telephone exchange
B   = Distribution cable
C   = Cordless base station
D   = Alterative positons for cordless sets
--- = Radio links

Figure 11.2 Home use of cordless telephone

handsets. But they are linked to their base units by radio rather than by wires (Figure 11.2). They contain batteries which have sufficient capacity to operate without recharging for about eight hours. They are automatically recharged when they are replaced on their base stations.

The cordless principle is also being applied to call-connect systems (Chapter 14) including PABXs. There is no reason why there should not be cordless non-voice terminals like PCs.

Cordless terminals could replace wired telephones, PCs and other terminals throughout a building. Many firms spend a great deal of money on moving telephones and terminals to keep up with changes in organization and office layout. Cordless terminals would save the great bulk of this. Coupled with modern short-distance radio systems they should make it possible largely to eliminate fixed wiring and cabling, not just within buildings but on complete complexes and industrial sites. Such thinking has led to the concept of the 'cordless PABX', with no extension wiring at all.

The latest development of cordless technique is the so-called 'telepoint' service. The principle of this service is that owners of cordless telephones would be able to take them to shopping precincts, railway stations, airports and so on and use them in conjunction with fixed transmitters and receivers at those locations. The effect would be to provide a service not far different from cellular mobile radio in those defined areas. Work on standards for such a service in Britain is complete and the first operations are due to begin in 1989.

Radio-paging

Radio-paging systems are already widely familiar. In a paging system, computer-controlled radio transmitters are connected to special numbers on the PSTN. Calls to these numbers cause selective radio signals to be broadcast, and these activate small receivers carried by individual users. When they are activated they emit tone signals which tell the user to call a prearranged telephone. Advanced pagers are now available which can display letters and figures — for example, the number to be called or a short message. Such systems have widespread application in developing countries.

Micro-cellular techniques

At the time of writing much attention is being given in advanced countries to the potential and development of micro-cellular systems. A micro-cellular system is similar in essentials to a cellular system, except that cell sizes and set powers are substantially smaller. The possible application of these techniques to provide personal communicators is being studied in Britain. The concept is of small, relatively cheap hand-held sets which could be used anywhere — inside and outside buildings. Such sets could not only supplement conventional cellular and other mobile services. They could also be used to provide a service which would compete directly with conventional wired telephone service, even in high-density areas.

One important consideration affecting the use of such radio systems should, however, be noted. The smaller the cell size the larger the number of radio base stations and supporting network links that will be needed. If high-density or blanket radio coverage is to be given over a significant area, like, say, a capital city and its suburbs, a very large investment may be required. It may just not be economic to make such an investment where even a poorly developed distribution cable network already exists.

Policy issues: rural and low-density distribution systems

Radio is therefore set alongside cable as the second principal transmission vehicle for communications. A basic difference between them is that in the case of cable the transmission medium itself — the copper wire or tube, or the optical fibre — has to be created and installed by human agency; whereas electro-magnetic radiation travels through the atmosphere (or outer space) on its own. The cost of radio is concentrated in the transmitting and receiving equipment. These characteristics mean that, as we have noted, terrestrial radio systems like micro-cellular radio have important potential for telecommunications customer distribution.

There are many practical constraints on the use of radio for this purpose in advanced countries, where distribution cable networks are almost universal and where customers are on average often only about a kilometre from the exchanges which serve them. Until recently, radio links were not competitive with ordinary cable for telecommunications requirements over distances less than about 15 km, except in unusually difficult terrain. But as the technology advances, as we have seen, serious consideration is now being given to the use of radio systems for static service over normal exchange-to-customer distances in Britain.

In developing countries conditions are much more suited to radio. Line lengths are often much longer; complete communities may be tens or hundreds of kilometres distant from the nearest point on the existing telecommunications network; and densities are low.

Radio systems are available now on the world market designed to be used to give service in such circumstances. A typical system can cover an area with a radius of 300 km and serve up to a thousand customers. It operates at 2 gHz and provides twenty channels.

Such systems may be the only way in which reasonably high-density telecommunication service can be provided in remote districts, since satellite facilities usually have to be confined to one or two circuits to a particular settlement and its environment, at most.

Wider merits of cable and radio in high-density areas

Even in high-density areas cables of any kind have some important disadvantages and radio systems have advantages which developing countries who are expanding their networks might do well to weigh.

From the technical point of view, telephone distribution cable networks used to be the most straightforward part of telecommunications. They consist essentially of large numbers of copper wires combined in cables and (in Britain) usually laid in ducts. The last few metres into the customer's premises often consist of a pole with individual pairs of wires radiating from it.

Although they are simple from the technical point of view, distribution cables of all kinds pose complicated practical and legal problems. Cables are clumsy to lay and maintain. They are always vulnerable to damage from excavations and so on. Work on cables is difficult to organize and supervise. Working conditions for the staff who install and maintain them are liable to be cramped and unpleasant. These staff face many hazards, for instance, gas accumulations and road traffic, which can be controlled but not eliminated. Cable excavations and reinstatements cause all sorts of hazards to other people.

There are other difficulties. In certain localities there is likely to be a steady increase in the requirement for broadband and other sophisticated distribution systems. If cables are used for these systems this will mean that increasing amounts of delicate electronic equipment are liable to have to be housed in manholes or roadside cabinets. However rugged the design and construction, such equipment will always be more at risk to damage than radio equipment, which can often be put in or on buildings.

From the point of view of efficient management of assets, a big disadvantage of cables and even more so of ducts is that if they turn out to have been put in the wrong place they have to stay there. It is rarely practical to re-lay a cable and never to re-lay duct. Even though the telephone service itself may soon approach saturation in certain respects, other telecommunications applications are still in their infancy and their distribution effects are as yet unforeseeable. There will always be fluctuation in distribution capacity requirements as buildings are put up, knocked down, put to new use and so on, and as businesses expand or contract. Demand can never be accurately forecast and the inflexibility of cable systems in this respect will always count against them.

Finally, the business of extending a wire or cable connection into a customer's premises to give or extend service is expensive, cumbersome and difficult to supervise and run efficiently. A radio system which enabled customers to buy handsets in retail shops, go home and have service commissioned by remote control from a central point would have great operational advantages to the enterprise.

The disadvantages of cable have been accepted in advanced countries in the past, mainly because radio systems were so much more expensive and spectrum was in such short supply. But as we have seen the position is changing. There is increasing discussion of micro-cellular radio and other techniques which would enable radio systems to be used with really high densities, comparable with those of conventional telephone service. Such systems are suited in principle either to be integrated into conventional wired service or to compete with it as 'personal communicator' systems.

There is a strong argument for this deriving from the changes in main telecommunications networks noted in Chapter 10. In modern networks all the other changes associated with providing a new customer line such as establishing the line's identity in stores associated with control and charging processors, actually connecting service and so on will be made by software changes from remote centres. In such circumstances it will be particularly anomalous that a physical visit still has to be paid to the customer's premises to install or alter wires and cables and sockets. It would be very much quicker, cheaper and generally a logical exploitation of the technology if the customer could simply buy a radio handset, notify his purchase and location over another telephone and have the whole process of connecting and commissioning service completed by the enterprise without a physical visit. A suitable customer radio system would provide this as a natural matter.

In developing countries circumstances are particularly favourable to such thinking. There is likely to be less pressure on spectrum. On the other hand, capital is often short and cable distribution systems are expensive. Radio systems have some important advantages because of their transferability. For example, service could be opened in an industrial or suburban area using radio at the start, while there is a low density of customers. The radio links could still be replaced by cable if desired as density increases and capital becomes available.

Existing customer radio systems are either essentially mobile, like cellular radio, or low-density 'luxury' facilities like cordless telephones. The issues involved in arranging substantive high-density customer radio service are complex and much detailed work still has to be done on them in the advanced countries. At the present time a wide variety of proposals are being made in these countries for applications of radio techniques to mobile and static service distribution of one kind or another.

In some of the most ambitious, visions have been held out of satellites servicing universal wrist radios. One may doubt whether these are practical in the foreseeable future, at least on a scale relevant to mass communications. Wrist radios that will work satisfactorily at frequencies of 12 or more gHz to satellites 23,000 miles away might conceivably be developed. But the practical difficulties of services of this kind are formidable and orbit spaces are limited. The use of satellites for any kind of widespread two-way communications distribution direct to the main mass of customers in most countries would pose extraordinarily complicated administrative and networking problems, at international as well as national level. It is doubtful whether this could be a practical prospect in the present state of technology and institutions.

On the other hand, advanced terrestrial radio distribution techniques like micro-cellular systems offer the chance of practical progress before long. There are many problems to be overcome, for example, in the development of fully satisfactory radio coverage and control techniques so that handsets can be used with confidence inside and outside buildings and anywhere in a country (or in the world); in the development of cheap handsets; and in the financing and provision of the necessary networks of base stations and supporting links. But the rewards for such developments would be great. Much effort is being devoted to such systems in various advanced countries at present.

It is clear that in all countries customer radio systems deserve careful and continuing consideration not only for mobile service but as a means of providing or competing with wired telephone service and ISDN connections as these become available.

Private systems

In advanced countries and in high-density areas in developing countries commercial forces have already stimulated the development of radio as a competitor to distribution cable and wiring on and to customers' premises in one important area.

This is for private networks and related fields. We noted that, traditionally, private networks have been made up of circuits rented in one or more local distribution cables, coupled up to longer-distance circuits as necessary. But the local cable network is laid out to suit the requirements of public-switched telephone service. It radiates from exchanges whose location is determined by switched telephone network economics. Compared with the shortest distance between the terminal points on private circuits such cables often follow very circuitous paths. Also, in many cases pairs in two or more cables have to be coupled together to create a private circuit and this takes time and costs money. It would obviously be impracticable to lay special cables in any but a few cases. For the great majority radio is the only serious alternative.

There is already radio equipment on the market which can be used to provide private analogue or digital links of considerable capacity in such cases. Over suitable terrain such radio links are becoming cost competitive with metallic cables over quite short distances. Radio technology and costs continue to improve, and we may expect radio to become more and more competitive for such applications.

Summary

To sum up:

— Radio techniques are well established as a means of conveying communications traffic between exchanges, for mobile and private customer systems, for broadcast distribution and for serving remote and difficult areas in developing countries.
— Techniques are being devised which would permit growing use of spectrum above 30 gHz. If this happens, spectrum problems at all frequencies should be eased.
— Cellular radio and cordless telephones are proving popular. There is every reason to think that cordless call-connect systems and PABXs will prove equally popular and valuable.
— The use of radio virtually to eliminate wires and cables at the customer end would be an attractive objective. It increasingly appears practical within customers' own premises and on individual sites; and micro-cellular radio techniques may make it possible to use radio to replace static telecommunications altogether in due course.
— Radio technology continues to advance and it is important that developing countries give careful and continuing attention to the use of radio systems for distribution in urban and other denser areas.

12 Planning, procurement and management of network plant

This chapter discusses the general principles of the planning, procurement and management of network plant. Generally speaking, the same principles apply to all networks — PSTN, Telex, Packet-switching, ISDN or IDN.

Network planning

The efficient planning of a network involves four questions:

— What kind of plant is to be used, in the technical sense?
— How much plant is to be provided to meet growth?
— How much is to be provided to replace existing plant?
— How is traffic to be routed within the network.

Technical choice and replacement policy

In the past exchanges have been the dominant element of networks from both economic and technical points of view — that is, what the industry calls 'switching'. The balance is changing as the result of technical developments and as the cost of exchange systems drops with the introduction of digital plant. But the issues involved are still well illustrated in the switching context.

Most of the new switching provided now in advanced countries is digital. All major manufacturers offer SPC digital exchanges as their principal product. So far as older systems are concerned, refurbished step-by-step (Strowger) is available from several sources on the world market. Crossbar and electronic analogue equipment based on reed relay switches developed in the 1960s and 1970s is still being manufactured. These older systems are normally used nowadays only for extension, as distinct from replacement, of existing installations.

In developing countries, however, so far as replacement is concerned, modern electronic analogue systems without SPC deserve serious consideration. They are considerably less demanding to maintain than SPC systems, mainly because software is not involved. It is true that their unit maintenance cost is higher than that of digital exchanges and that they can only provide some of the more sophisticated modern facilities, like ring-when-free, if they are supplemented by add-on units. But in the shorter term and in the circumstances of most developing countries this is of marginal importance, compared with basic matters like maintainability.

BPOT faced a major switching choice in 1972 (Whyte *et al.*, 1974/75). At that time two decisions had to be made:

— whether to replace the existing population of Strowger (step-by-step) exchanges with modern plant;
— if they were to be replaced, whether to use SPC crossbar or electronic analogue equipment.

Questions of this kind are very complex and involve a large number of economic variables as well as strictly technical questions. In fact, most technical issues can be reduced to economic terms. For example, the technical maintainability of two different systems can be expressed by estimates of what it would cost to maintain them for a given standard of service; and thus its significance can be directly related to that of capital cost.

In BPOT techniques were devised for addressing these issues by means of a computer model. A large number of variables were carefully evaluated for each of the three alternatives. They included capital cost,* and annual cost including maintenance, power, accommodation and other charges. The capacity of the industry to supply was agreed with the appropriate firms and included as a constraint. The model also required forecasts of demand variables, like number of lines, number of calls, average call duration and so on. Each input was as thoroughly checked as it could be by the specialists concerned in BPOT and by discussion with the supplying industry.

The procedures took account of the benefits of new facilities. They also allowed the wider social benefit implications of the various approaches to

---

* The capital cost of an exchange can conveniently be expressed as a polynomial expression of the form $ax + by + cz$, where a, b and c are amounts of money and $x =$ number of lines, $y =$ traffic in Erlangs and z is a fixed cost per exchange to cover power and similar plant. In the BPOT case described values for a, b and c were developed in discussion with the suppliers concerned. The Erlang is the standard measure of telecommunications traffic load or occupancy. It is most simply understood as the volume of traffic which would occupy one circuit continuously throughout the busiest hour of the day.

be assessed. The model was run. The cash flows were evaluated by the Discounted Cash Flow (DCF) technique using an interest rate specified by central government. The financial results obtained were used as the basis of major submissions to the (then) Post Office Board. The Board was able to take a solidly based and well-informed decision on the two issues, uninfluenced by supplier pressures. The decision was ratified by Government. It was decided to proceed with a programme to replace all Strowger exchanges by 1992 — that is, over roughly a twenty-year period — and to use the TXE4 and TXE4A analogue electronic systems, with the proviso that they would be superseded by digital equipment when this was available and economic. Such a programme was shown to have a limited economic advantage; given this, it was regarded as justified primarily by the deteriorating service to be expected if Strowger exchanges were retained indefinitely. The model and the associated procedures proved so satisfactory that they were used again when the decision to adopt digital switching in Britain came to be taken.

The full procedure used by BPOT requires a great deal of the time of expert staff and extensive computing procedures. A special group was formed simply to carry out the work. But simpler procedures, capable of being devised and used in many developing countries, should suffice to allow sensible decisions to be taken. A model which incorporates capital and maintenance costs could often be used first to give a broad indication. If the results are decisive in favour of a particular option, this may be all that is needed. If not, a fuller model more like that used in Britain may be needed.

Similar questions of growth and renewal arise for transmission plant, except that the technical and economic choices are usually less complex. A similar approach is appropriate. Local cable planning involves special problems discussed below. For a discussion of more recent questions of network strategy in BT, see Shurrock and Davis (1982-3).

Dimensioning and routing

In analogue systems the correct matching of exchange and transmission plant to predicted customer traffic and leased circuit requirements and the optimization of the engineering economics of networks are important matters. Both surpluses and shortages of lines or traffic capacity are serious. Shortages mean that the network is failing to do its job in the economy and the community, and that revenue is being lost. Surpluses mean that an undue burden of idle investment is being carried, with serious implications for the finances of the enterprise.

In processor-controlled systems it is just as important to match the physical capacity of the network to load overall, even though the controls may be able to match moment-by-moment flows to capacity by varying routings across the network. The approach to such matching is discussed in more detail below.

All plant projects involve the determination of the quantity of equipment to be provided. This process is called dimensioning. A forecast of demand — lines, calls and load in Erlangs is required; the optimum routing pattern for traffic between exchanges has then to be determined; and equipment quantities and cable and radio link sizes then have to be decided.

In many advanced countries it is now accepted that both exchange area layouts, which will have been based on assumptions about the economics of electro-mechanical or even manual plant, and the pattern of network functions — for example, whether processing should be divorced from call connection and carried out at special centres — should be reviewed when major schemes are being planned using SPC digital equipment. Such questions will clearly need to be looked at in some developing country situations using similar appraisal techniques to those described above. But the issues involved are highly technical and outside the scope of this book. In some developing countries it may be best to address them with outside help — for example, from ITU TCD.

Virtually every element of older electro-mechanical exchanges, particularly those using Strowger (step-by-step) equipment, can be closely tailored to need. Quite sophisticated procedures were developed over the years in BPOT to do this. These procedures are outlined below. It is suggested that it would be desirable for developing countries who still have electro-mechanical plant in their networks to apply similar procedures.

Some elements of digital systems are supplied in relatively large modules of capacity. A digital switch is likely to have capacity to handle many simultaneous calls. On the other hand, costs have been reduced so much that the unit cost of the plant per call tends to be much lower than the equivalent for electro-mechanical plant. Close tailoring of these elements to avoid surpluses is therefore neither necessary nor practicable although shortages are as serious as ever. On the other hand, some elements of digital exchanges are still provided on a one-per-customer basis — for example, customer line terminations. A considerable part of the cost of the exchange is tied up in these. The procedures described below are appropriate to be applied to them.

The determination of the optimum routing of traffic between exchanges (or in modern processor-controlled networks between nodes — that is, switching points or multiplexers) is an important activity in its own right. It involves the application of formulae derived from the economics of the plant to forecast traffic patterns, to determine the paths calls should follow and whether they pass directly between local exchanges or through intermediate switching points. Especially in processor-controlled systems the detailed procedures are complicated and require expert technical attention.

In BPOT it was standard practice for dimensioning and routing to be determined by BPOT staff. A detailed specification showing plant quantities was then passed to the supplier(s) concerned. BPOT experience was that if suppliers were left to dimension projects for themselves, they were liable to under-provide as well as to over-provide (which might seem more natural).

Forecasts used in plant planning

Plant dimensioning and the preparation of routing plans calls for exchange-by-exchange and route-by-route (or node-by-node and link-by-link) forecasts of lines, calls, load in Erlangs and various other parameters like the average duration of calls. For a discussion of techniques in these respects see Holmes (1974/75) and Farr (1988).

Such forecasts are best made as close to the ground as possible. But it is good practice to confirm that in aggregate they reconcile with the similar forecasts for the enterprise as a whole compiled at national level. In BPOT this was achieved by a 'bottom-up, top-down' process of debate between field units and Headquarters carried out each year as part of the general planning cycle (See Chapter 5).

These forecasts are best prepared by staff independent of those responsible for calculating equipment quantities or placing contracts. Very few forecasts are exactly right. In principle it is desirable for forecasters to aim at a forecast which is as likely to be high as low. It follows that procedures for converting forecasts into equipment quantities and for management and redeployment of capacity in service must allow for the fact that any given forecast may be too high or too low — a point to which we shall return.

Forecasts of customer lines required are used to provide local distribution cables as well as parts of the exchange. They therefore need to be broken down into small units like streets or housing or industrial estates which can be related to cable layout. In BPOT the forecasters assembled as much local knowledge as they could about building and other developments in the area from outside sources like planning authorities, state and commercial developers and so on. They then studied the situation on the ground and prepared a forecast in a format broken down to cater for the needs of the local cable planning engineers, with whom they were in close touch. Forecasts for exchanges were readily assembled by aggregating the forecasts prepared for local cable purposes.

Forecasts of calls originating from customers and of traffic load in Erlangs involve different techniques. The first step is study and interpretation of past records of these variables. Erlangs are in effect the product of the number of calls and their average length or duration. Voice durations do not normally change much over time (except when there is a major change in charging pattern such as that which occurred in Britain when periodic metering was introduced; see Chapter 6). It is therefore usually an acceptable compromise to study records of the number of calls and to use these to deduce the trend of growth in Erlangs. But voice durations should still be checked by sampling from time to time. Non-voice durations may be markedly different from those for voice traffic and special attention may be needed to make sure they are correctly catered for.

Past growth in the total number of calls (local, long-distance and international combined) on an exchange will usually be found to follow a similar trend to growth in lines, except for one factor. Additional business lines usually carry as much traffic as existing ones, so that traffic on a purely business exchange often grows closely in line with the number of lines. Additional residential lines, on the other hand, tend to carry fewer calls than business lines, so that the more residential lines are added the lower the average number of calls per line (or calling rate) will be. In such circumstances the rate of growth of calls may be lower than the rate of growth of lines. In Britain this effect was compounded as lines were added in lower socio-economic categories, because in Britain customers in these categories tend to make fewer calls than those who are better off. On the other hand, in the early stages of development of a network customers in outlying areas often tend to make heavier use of the telephone than those close to urban centres.

In advanced countries such as Britain it has been found that when telephone density reaches a certain level the growing usefulness of the telephone to residential customers because of the number of their contacts who have service outweighs this process of dilution. At this point the calling rate starts to rise again. In most developing countries this effect is probably some way off yet. But it clearly needs to be watched for. Both effects can be seen in Figure A4.5, Appendix 4, which plots the behaviour of overall inland calling rate in the United Kingdom.

The rates of growth of long-distance calls and of international calls require to be forecast separately for planning long-distance and international facilities. In Britain the rate of growth of long-distance (trunk) traffic is usually higher than the rate of growth of total or local calls.

It is stated in Chapter 6 that demand for telecommunications service is inelastic to price except in the very short term. This assumption may not of course hold good in particular circumstances in developing countries. It can usually be tested by examining past records for growth in calls and lines and relating them to tariff changes. Clearly if lasting effects are found allowance should be made for known future tariff increases in project forecasts. In practice this is not likely to be easy. Proposals for price rises are usually politically sensitive and it may not be possible to make them known in time to people concerned with project design. In such circumstances project forecasts may have to be specially adjusted centrally before they are used.

Call forecasts prepared from interpretation of past records in this way should be related to expected future developments like expected economic trends in the area concerned, major development schemes like factories, dams and so on. In the final analysis all forecasts involve a high degree of judgement. Large sums of money have to be committed on the basis of project forecasts and it is good practice to require the forecasts to be approved at senior level before they are used for engineering design.

Influence of different factors

The dimensioning process involves not only forecasts but a number of other inputs. BPOT experience was that the practical significance of these varied. In ascending order of importance the most significant are:

1. 'Grade of service' standards which specify the number of calls which are permitted to fail in the busy hour of the day due to limitations on the capacity of the plant. Such standards are essential, because it would be quite uneconomic to plan to carry all the traffic which is offered. Values for these standards have been established empirically within the industry over the years and are a well-understood element of engineering technique. A typical standard would be that not more than one call in 200 is to be planned to be lost in the busy hour in a stage of electro-mechanical switching. Contrary to what might be expected, practical variations in these standards do not have a major influence on plant quantities in projects.
2. Equipment modules — for example, exchange capacity for line terminations — might be provided in blocks of 200. In most cases some mismatch between precise requirements and the nearest module is of course inevitable, and in practice judgements have to be made about this in almost every part of an exchange or transmission repeater station or cable network.
3. Design periods — that is, the number of years ahead for which plant is provided. This period is made up of (a) the time needed to design and procure the exchange; (b) the time for manufacture and installation; (c) a 'provisioning' period determined by management decision (in BPOT after careful economic study) for the number of years the tranche of equipment concerned is to be planned to last (typically three years for equipment in an electro-mechanical exchange).
4. The inevitable error in forecasts.
5. Supplier delivery performance — that is, the difference between the manufacturing and installation time assumed in 3(b) and that achieved in practice. For a variety of reasons BPOT has had extensive experience of such delays. They can be a major cause of waiting lists and congestion.

Senior management attention should probably be concentrated on the later items in this list, especially 3, 4 and 5.

In BPOT the collective influence of all these factors in creating surpluses and shortages in the network and thus on service and economic performance was found to be considerable. Regular disciplines were therefore instituted for monitoring the situation on the ground. Asset Utilization Factors (AUFs) were derived by measuring traffic flow in Erlangs and counting the number of lines; and relating the results to installed capacity at the specified Grade of Service. Target values were calculated for AUFs having regard to the spare capacity required by the provisioning periods. Measured and predicted AUFs for up to five years ahead were derived from measurements on the ground and from additional capacity currently on order or planned. They were discussed with field line managers as part of the regular budgetary and planning cycle.

In electro-mechanical exchanges measurement of traffic flow is a routine task usually allotted to the general maintenance staff. It is liable to be neglected because of the more pressing needs of maintenance proper. This should not be allowed to happen. The importance of regular traffic measurement and of management monitoring of the results cannot be overstressed. It provides an essential basis not only for AUFs and similar controls but for the detection and rectification of shortages (and therefore bottlenecks) and surpluses on the ground, and for the planning of new exchanges and extensions. In SPC exchanges measurements can normally be carried out automatically and the results can be processed on central computers connected on-line.

Traffic patterns change all the time. In a growing network it is essential that load on individual inter-exchange or inter-node routes is checked at regular intervals. In BPOT a regular system of 'Annual Schedules of Circuit Estimates' (ASCEs) was developed in which five-year forecasts of traffic and circuit requirements on each route were updated every year. Plant provision programmes were adjusted accordingly. Attention should be drawn to the importance of making sure that such estimates are based on up-to-date measurements. Serious errors can be introduced in a growing network if forecasts of this kind are not based on correct information about present traffic.

As already noted, such requirements have greatest force for electro-mechanical analogue switching and transmission plant, where circuits are commissioned and switches may be added individually. In digital transmission systems circuits are usually commissioned in blocks of at least twenty-four (North American standard) or thirty (CCITT standard) and switching capacity in larger modules, so that close tailoring is inappropriate. Nevertheless network economics and service performance can still be seriously affected if load and capacity get out of step. It is particularly important to avoid shortages or, equally important, gross over provision for all kinds of plant.

Local line planning

The basic approach to local line (distribution cable) planning is similar. Provisioning periods are usually longer because it is not practical to revisit sections of the cable network as often as exchanges. The determination of the exact period (typically seven years) is a matter for careful economic study.

Line plant planning is a difficult process. The detailed configuration of local cable or overhead wire networks and the choice of cable sizes and duct and pole route layout are a matter for properly trained engineers. The work is of such a character that it has to be carried out by relatively large numbers of staff. It is usual to employ staff of groups 2 and 3 in Chapter 8. Thorough and experienced supervision is most important.

Recent studies in BT have underlined that there is only a limited increase in the capital cost of schemes if local line plant is designed with generous margins of capacity; that the money saved by unduly close tailoring of provision to forecast need may be more than offset by the later overall cost to the enterprise of shortages leading to waiting lists for service; also that unless the main body of the network is in good condition economies in line plant provision may be outweighed by the saving in maintenance from having sound pairs freely available.

It will always be essential, however, to make sure that a balance is struck between such considerations and seriously uneconomic over-provision, and also that so far as humanly possible ducts and cables are put in the right place. Cables in land which is undeveloped and turns out to remain so, or in areas where there is little demand for telecommunications are valueless.

In the past local line planning has not had the attention it deserves, partly because it did not have the prestige of high-technology plant. This is changing as high-technology such as optical fibres and sophisticated electronics are introduced into the distribution network. But however advanced the technology of the associated electronics, proper planning and physical construction of the local cable network are essential. Senior management should ensure that it gets the necessary attention.

Building planning

An enterprise which is expanding its network is likely to be involved in a substantial amount of building construction and modification. Electro-mechanical exchange equipment and analogue transmission equipment usually require specially strengthened floors and higher ceilings than normal. Digital plant can usually be accommodated in conventional buildings. Also it generally requires less space than analogue equipment. Buildings erected for analogue equipment which have spare space can therefore usually accommodate much more capacity in digital plant than the analogue capacity for which they were designed. The result is that enterprises are usually able substantially to reduce the size and number of new building schemes once it is decided to introduce digital plant.

Substantial investment is involved in buildings and shortages of building capacity take a long time to deal with. Buildings should be carefully and methodically planned by professional engineers and architects. New buildings and extensions are usually designed to last much longer — that is, to absorb much more growth — than tranches of exchange equipment. BPOT designed its exchange buildings to last between thirteen and twenty years, depending on the economics of the particular case. Building sizes can be determined by preparing estimates based on the same kind of forecasts as are described above for plant planning. It is important that building designs take account of the eventual layout of the equipment and the processes needed to install it. A great deal of money can be wasted if ports for cables or loading doors for the entry of equipment are put in the wrong place.

It is important that the successive stages in a construction project are properly coordinated. Building construction will usually start first, followed by procurement and manufacture of equipment. It is essential that the building is fully ready for installation when the equipment is ready to be delivered. It is standard practice in BT to use Critical Path Analysis (PERT) techniques to coordinate projects in this way.

Short-term expedients

As we have noted, forecasts are more often than not wrong and supplier delays will occur from time to time. Temporary conditions of shortage of exchange equipment or local cable pairs are bound to develop from time to time unless plant is extravagantly over-provided. Mobile exchange equipment mounted in trailers is available from various sources on the world market and can help greatly with exchange shortages. Local line shortages are often more difficult to deal with, since a cable in the wrong place cannot often be moved and new ones take time to provide. Various expedients are, however, available. The most important are radio systems, which can be much more quickly installed than cables, and concentrators, which typically allow calls from ten lines to be carried on three cable pairs. Skilled use of such expedients can do a lot to contain the effects of temporary shortages.

Procurement of major network plant

Telecommunications is a capital-intensive business, and the characteristics and performance of its capital plant are of fundamental importance to a telecommunications enterprise. Very large sums of money are often involved in orders for telecommunications plant. Procurement therefore requires very close attention and care in organization (The procedures described here are appropriate for network plant and mainframe computers. Procurement of customer apparatus is dealt with in Chapter 14.) This section summarizes the conclusions to be drawn from BT experience.

Procurement procedures — that is, the administration of tendering, the negotiation and award of contracts and related commercial matters — should be carried out by a specialist group organizationally separate from the rest of the enterprise and from engineering groups in particular. It is, however, important that this group is physically located close to the engineers with whom it has to deal. Its activities should be under close supervision by senior management, and subject to regular audit. The staff concerned should be aware that they are subject to audit.

The aim of procurement operations should be to secure a correct balance between low initial cost, reliability, maintenance, power and other costs in service and acceptability and dependability of suppliers' offers of delivery and completion dates. Engineering views should be sought as a matter of course and given due weight, but the responsibility for procurement decisions should rest clearly with the procurement group. Decisions about award of major contracts should be submitted to very senior management.

There should be close liaison between procurement and engineering groups to ensure the best results. If tenders depart from specification careful judgements have to be made about the extent to which the departures are acceptable.

To the greatest possible extent procurement should be conducted by competitive tendering against published criteria — that is, price, quality, completion date, physical size, facilities, power consumption and so on. Decisions to award orders to one supplier without competition should be subject to particularly stringent checks. If, as is often the case, the grounds for non-competitive procurement are technical or operational (for example, that no other supplier can meet the required delivery date), they should be formally certified in writing by the technical or operational group concerned. Norms should be set for reasonable profit levels for non-competitive orders and steps taken to enforce them — for example, by requiring suppliers to open their books. Such a norm might be expressed as a permissible return on capital to be earnt by the supplier on the facilities involved in executing the order.

Telecommunications plant is usually procured using closed tendering procedures — that is, tenders are not advertised but are sought from a small number of suppliers known to be able to carry out the work. (Changes may be made in this respect in European countries soon at the instance of the European Commission.) Decisions about which firms to approach are often as important as decisions on the eventual award of contracts. They require close attention and oversight from senior management.

Procurement is difficult work. It calls for high-calibre staff with unquestioned integrity and a clear sense of when to consult their superiors.

Technical authority

For many years BPOT was the principal technical authority within the telecommunications industry in Britain. It maintained a large development staff and was accustomed to write detailed product specifications for its requirements. In other advanced countries arrangements have always been different. The administration has specified its requirements only in terms of the details of performance required from the equipment, and design and development has been carried out within the supplying industry. Even in BT the practice has changed in recent years, partly because of changes in the technology and partly as a matter of policy.

Very large developing countries may decide to build up substantial technical capability within their operating enterprises for broader policy reasons. For the generality of developing countries this may not be appropriate. But it is important that all enterprises of any size develop sufficient technical competence to be able to scrutinize suppliers proposals effectively and to be able to apply quality assurance and other checks on plant before it is accepted. Unsatisfactory plant should never be accepted or paid for.

Summary

To sum up:

— Since the early 1970s BPOT has used carefully validated computer-based appraisals for major decisions about choice of plant and technology.
— Surpluses and shortages of plant have major implications for service and economics. In the mid-1970s BPOT developed disciplines for controlling them. These questions remain important even if advanced technology is used. It is important for senior management to give attention to them and to the forecasts used in all sectors of plant planning.
— Local cable planning is a particularly difficult sector with great influence on service; it deserves special attention.
— Procurement practices require high-calibre staff, close senior management attention and regular audit. Wherever possible procurement should be by competition.
— Developing country enterprises should have sufficient technical expertise of their own to be able to scrutinize suppliers' proposals and to apply proper quality and acceptance tests.

13 International telecommunications and the ITU

The telecommunications industry has always given high priority to progress in international telecommunications. They have become a separate and major sector. This chapter includes a brief description of the institutions that regulate international telecommunications at world level and of the technology and capabilities of international links and the world network.

International communications play a major part in the economic activity of most countries. The importance to developing countries and their economies both of good international links and of good supporting networks serving major business and industrial centres within the developing countries concerned cannot be overstressed.

The International Telecommunications Union

The principal international agency concerned with telecommunications between countries is the International Telecommunication Union (the ITU). Its headquarters are in Geneva. It is a specialized agency of the United Nations. The members of the ITU are Governments. For example, Britain's formal representatives in ITU are Government officials. From time to time ministers may be involved. BT, other operators such as Mercury, and in some cases representatives of manufacturing industry participate extensively in ITU activities at working level.

The purposes of the ITU as defined in its Convention are:

— to maintain and extend international cooperation for the improvement and rational use of telecommunications of all kinds;
— to promote the development of technical facilities and their most efficient operation with a view to improving the efficiency of telecommunications services, increasing their usefulness and making them, so far as possible, generally available to the public;
— to harmonize the actions of nations in the attainment of those common ends.

The ITU works to fulfil these basic purposes in three main ways:

— international conferences and meetings;
— technical cooperation;
— publication of information, world exhibitions and similar activities.

Plenipotentiary Conferences

The Members of the Union meet at a Plenipotentiary Conference at intervals of normally not less than five years. The Conference is the supreme authority of the ITU and lays down general policy. It reviews the Union's work since the last conference and revises the Convention if it considers this necessary. It also establishes the basis for the organization's budget and sets a limit on expenditure until the next conference. Finally, it elects the Members of the Union who are to serve on the Administrative Council, the Secretary-General and the Deputy Secretary-General, the members of the International Frequency Registration Board (IFRB), and the Directors of the International Consultative Committees (CCIR and CCITT) (see below). The elected officials hold office until the following conference.

Administrative Conferences

There are two kinds of Administrative Conference held by the Members of the Union: World Administrative Conferences and Regional Administrative Conferences. The agenda of a World Administrative Conference may include: the partial revision of the Administrative Regulations (Telegraph Regulations, Telephone Regulations, Radio Regulations, Additional Radio Regulations), the documents which govern the international operation of the three modes of communication; exceptionally the complete revision of one or more of these Regulations, and any other question of a world-wide character within the competence of the conference.

The agenda of a regional administrative conference may provide only for specific telecommunication questions of a regional nature, including instructions to the International Frequency Registration Board (IFRB) regarding its activities in respect of the region concerned, provided such instructions do not conflict with the interests of other regions. The decisions of such a conference must in all circumstances be in conformity with the provisions of the Administrative Regulations.

Administrative Council

The Administrative Council is composed of forty-one Members of the Union elected by the Plenipotentiary Conference. It normally meets for about three weeks once a year at Union Headquarters in Geneva and at these formal sessions acts for the Plenipotentiary Conference between the latter's meetings. The Council supervises the administrative functions and coordinates the activities of the four permanent organs at ITU headquarters and examines and approves the annual budget.

The International Frequency Registration Board (IFRB)

The IFRB consists of five independent radio experts all from different regions of the world, elected by Plenipotentiary Conferences and working full-time at the Union's Headquarters in Geneva. They elect a Chairman and a Vice-Chairman for each year from among their own members.

The Board's main task is to decide whether radio frequencies which countries assign to their radio stations (and which they have notified to the Board) are in accordance with the Convention and the Radio Regulations and will not cause harmful interference to other stations. If the Board's finding in a particular case is favourable, the frequency is recorded in the Master International Frequency Register kept by the IFRB and thus obtains formal international recognition and protection. An average of more than 1,200 frequency assignment notices, covering new assignments or changes to existing assignments, arrive at the IFRB each week.

Among the other major tasks of the IFRB are: participation at the request of governments in the obligatory inter-governmental coordination of the use of frequencies involving space techniques prior to their notification for recording in the Master Register; the orderly recording of the positions assigned by countries to geostationary satellites with a view to ensuring their formal international recognition and the technical preparation of radio conferences with a view to reducing their duration.

There are two Consultative Committees — the International Radio Consultative Committee (CCIR) and the International Telegraph and Telephone Consultative Committee (CCITT). The two CCIs are separate organs of the ITU dealing respectively with radio communications on the one hand and telegraph, telematic and telephone technique, cooperation, tariff principles and international accounting on the other. All Member countries of the ITU are also members of right of the CCIs with their Recognized Private Operating Agencies (RPOAs — such as BT and Mercury) if requested; manufacturers, scientific and international organizations may also join CCITT work in those Study Groups they are interested in.

Each CCI holds a Plenary Assembly every few years. The Plenary Assembly draws up a list of technical telecommunication subjects or 'Questions', the study of which would lead to improvements in international radio communication or international telegraphy and telephony.

These Questions are then entrusted to a number of Study Groups, composed of experts from different countries. The Study Groups draw up Recommendations which are submitted to the next Plenary Assembly. If the Assembly adopts the Recommendations, they are published. CCIR and CCITT Recommendations have an important influence among telecommunication scientists and technicians, operating administrations and companies, manufacturers and designers of equipment throughout the world.

Technical cooperation

The Technical Cooperation Department of the General Secretariat, mainly within the framework of the United Nations Development Programme (UNDP), administers a programme through which telecommunications experts are sent to various countries throughout the world to advise on the operation of all terrestrial and space telecommunications systems or to help train technicians of the future. In addition, there are many students and fellows studying telecommunications under this programme in countries other than their own. Modern regional telecommunication networks have been set up or are in the process of being set up in Latin America, Africa (PANAFTEL), Asia and the Mediterranean and Middle East Region (MEDARABATEL).

The IFRB provides technical advice to Members of the Union to enable them to operate effectively as many radio channels as possible in the overcrowded parts of the radio spectrum where there is liable to be harmful interference between stations. In addition, the IFRB investigates cases of harmful interference reported to it and makes recommendations to the countries concerned on how best to solve their particular problem.

CCITT technical assistance activities, including for instance the preparation of handbooks drawn up by the Special Autonomous Groups (GAS) and certain Study Groups, are conducted with extreme care, to ensure that the target administrations derive maximum benefit. The handbooks are not restricted solely to the technical or economic aspects of telecommunications, but also cover problems of planning (including rural telecommunications) and new technologies (digital network).

Over recent study periods, the activities of the (World and Regional) Plan Committees have expanded considerably, thanks to the active participation of a large number of administrations. On the one hand, the Plan Committees prepare general (world and regional) plans containing existing data and forecasts, processed by computer, providing a database for the planning of telecommunications networks. Secondly, in pursuance of a Resolution of the CCITT Plenary Assembly, during the Plan Committee meetings technical lectures have been given by eminent international specialists from CCITT Members, followed by discussions, in particular on studies in areas which have a direct impact on planning and decision-making for the development of national and international networks.

Centre for Telecommunications Development

The Centre for Telecommunications Development came into being on the recommendation of the Maitland Commission (Appendix 1, Paragraph 9).

The Administrative Council of the ITU decided to establish the Centre in July 1985. It constituted the first Advisory Board — representing different regions and interests — to guide the Centre, mobilize resources, and approve its programme and budget.

The objectives of the Centre as defined in its mandate are:

— to stress and promote the key role of telecommunications for socio-economic development so as to enhance investments in this sector;
— to provide advisory services as well as operational and technical support in critical areas so as to foster telecommunications growth in developing countries;
— to raise and mobilize resources, to initiate cooperative activities, and to provide information services so as to optimize the overall effectiveness of telecommunications development.

The Centre's first Action Plan (1987–89) was approved in April 1987. Its implementation is closely coordinated with the complementary activities of the Technical Cooperation Department (TCD).

Communication between countries

International telecommunications involve technical and commercial arrangements between the operators in the countries concerned. In the nature of things these arrangements tend to be complex and specialized. Standards and procedures for them are agreed and promulgated through CCITT. The precise arrangements which are to apply between two countries are then settled bilaterally between the operators concerned.

For example, in the technical field a series of specifications for signalling between international exchanges has been promulgated over the years by CCITT. An important recent one, CCITT No. 7, prescribes standards for signalling between exchange processors in SPC exchanges. On any given route, it is a matter for agreement between the specialists of the operators concerned whether inter-exchange signalling, say, between London and Paris, conforms to a CCITT specification or is based on a national or proprietary specification. Again, CCITT rules specify the procedures for accounting arrangements between operators, and that settlements are to be made in Gold Francs. But the actual charging and accounting arrangements on calls between two countries' operators' customers and the determination of any cash settlements between them are a bilateral matter for the operators concerned.

Intercontinental transmission

For a long time the most challenging problem in international telecommunications was a technical one. It has been possible to send telegrams between countries and continents for well over 100 years. The first reliable submarine telegraph service between London and Paris began to operate in 1852. The first cable between Britain and North America began to operate in 1866. But these cables could not carry speech.

Telephone calls could be circulated between countries with common land frontiers using the techniques used for domestic telecommunications. But the creation of telephone links across oceans presented very difficult technical problems.

In the 1920s short-wave radio (Chapter 11) began to be used to provide intercontinental telephone circuits. Radio telephone service between Britain and the United States opened in 1927, for example. But by modern standards the number of circuits available was severely limited. Radio calls could only be made at certain times of day on any given frequency, and they were subject to fading and interference of many kinds.

Submarine telephone cables

Special attention was therefore given to the development of submarine telephone cables. The submarine telegraph cables laid in the nineteenth and early twentieth centuries usually consisted of a single copper conductor enclosed in a sheath of gutta-percha insulation with steel or similar metal cladding. Such cables could only convey speech for a few miles before it became too weak to be useful. The problem was to develop suitable cables and submarine amplifiers or repeaters which could be inserted at intervals along the cable to restore the strength of speech.

The design of such repeaters presented severe technical challenges. They had to be capable of withstanding the stresses of being laid with the cable and of resisting the great pressure of water at oceanic depths. They needed to be fed with electricity over distances up to thousands of miles. The only practical way to do this was along the cable. The repeaters had to be extremely reliable since it is very difficult and expensive to bring a submarine cable back to the surface once it is laid.

These problems were solved by a series of important advances in mechanical, marine and electronic engineering. Submarine telephone cables were developed which used low-loss coaxial construction. Repeaters were developed for association with them which used extremely reliable valves. They could be laid and operate reliably on the bottom of the world's major oceans.

The first experimental submarine repeater in the world was laid by BPOT in the Irish Sea in 1943. The first pair of transatlantic telephone cables (one for each direction of transmission) came into service in 1957. From then on, 24-hour telephone service of normal quality was available between Europe and North America. Submarine telephone cables were soon laid across the other major oceans. By now all the major continents and subcontinents and many major islands are linked into the world telecommunications network by cable. The length and capacity of submarine cable systems has increased steadily over the years.

Communications satellites

We looked briefly at communications satellites from the technical point of view in Chapter 11. From the outset it was realized that they could become an important medium for international telecommunications.

The design of commercial communications satellites involved just as many technical problems as that of submarine telephone cables. They had to operate extremely reliably and be fed with electric power in an even more hostile environment than the ocean bed. They required large, expensive earth stations which had to meet extremely demanding requirements for sensitivity, precision of movement and stability. They presented special commercial and political problems. They required very specialized and expensive launching facilities, and they had a useful life of less than ten years.

All these problems were successfully overcome. An international organization — INTELSAT — was brought into being by interested governments to oversee world-wide communications satellites and related matters.

Satellite systems have one inherent limitation. The distance from the surface of the Earth to the orbit used by geostationary telecommunications satellites is approximately 23,000 miles. Communications have to travel twice this distance if one satellite is used. Even at the speed of radio waves, which is that of light, this distance introduces an appreciable delay, of about 800 milliseconds (thousandths of a second). This is tolerable on computer traffic, and just about tolerable on voice traffic. But the delay makes it impossible to use two satellites in tandem on voice calls.

Optical-fibre submarine cables

Satellites and metallic submarine telephone cables have been commercially competitive with one another for some time as transmission media for transoceanic distances. Optical-fibre cables are very suitable for submarine telecommunications systems. They have greater capacity than metallic cables. The very rapid advances in optical-fibre technology constantly reduce the number of repeaters required for a given distance. In theory technology now exists which would make it possible to lay a transoceanic telecommunications cable which requires no submerged repeaters at all. The advantages of such a cable, in terms of ease and speed of laying, initial and operating cost and reliability will be very significant.

It seems possible that in time such cables will largely supersede both satellites and metallic cables for high-density inter-continental telecommunications, for example, across the North Atlantic. But satellites will certainly have an important role in providing international links for developing countries for many years to come.

International telephone exchanges

Domestic telecommunications networks usually incorporate international exchanges which are used to switch traffic to and from other countries. These exchanges have to intercommunicate with the corresponding exchanges in other countries throughout the world. Their inter-exchange signalling equipment or software is therefore specialized and diverse. They must also have facilities for recording details of the calls customers make to all the foreign countries to which they are connected, as the basis for the settlement of accounts between the various national administrations. The requirements for routing calls to and through foreign administrations' networks are more complex than those for purely domestic traffic.

Apart from these specialized features, however, international exchanges are much the same in all their essentials as inland long-distance exchanges. Customers are able to dial international calls in the same way as they dial domestic trunk calls.

The generations of international exchange equipment correspond closely to those of domestic exchange equipment. In Britain, purely electro-mechanical international automatic exchanges continued to be installed up to the late 1970s. Digital electronic exchanges have come into use in the 1980s for international traffic, just as they have for domestic traffic. BT has separate international telex and packet-switching networks and exchanges, arranged in much the same way as their domestic counterparts.

In recent years the world inter-continental telecommunications network has expanded to meet demand in the same way as the domestic networks of the countries it links. Every kind of communication — speech, telex and all kinds of data, still and moving pictures, and full-colour television — which can be carried on domestic or continental networks can now be carried on inter-continental links. International traffic has grown, and continues to grow, at impressive rates throughout the world, not least to developing countries. As the cost and reliability of international telecommunications continue to improve, and the range of facilities available continues to expand, high growth will no doubt continue.

The great bulk of the world's long-distance international links have been provided since 1960. The plant in them is usually modern and has been designed with special attention to reliability. Most international exchanges are also modern. As a result the international network is more reliable, and less subject to interference from clicks and bangs than many of the national networks it connects with. The capability of an international switched telephone connection to handle data transmission and comparable services is usually governed by the state and capability of the networks in the countries at the two ends. For example, the international telephone network can carry bit rates which usually at least match those that national networks can carry. At present international switched telephone circuits can be expected to carry up to 9.6 kbit/s with less difficulty than their analogue inland equivalents. As digital plant penetrates international exchanges and links, they will be able to transport non-voice traffic at speeds up to 64 kbit/s per channel.

A number of countries now have international packet-switching exchanges which can exchange packet traffic between national packet networks. We noted in Chapter 10 that the advantages of packet-switching service compared with circuit-switching increase with distance. The international packet network is an important vehicle for non-voice communications.

Many multinational companies and other large organizations rent international private circuits and international private networks. Many multinational companies now manage and conduct world-wide operations using computers and many kinds of non-voice terminals linked by their own intercontinental private networks.

Summary

To sum up:

— Good international links and good supporting networks serving business and industrial centres are vital to the economies of developing countries.
— the technical problems of inter-country and inter-continental communications have been overcome. Further technical advances may be expected to concentrate on reducing costs and improving capacity and reliability.
— International exchanges and lines are predominantly made up of modern plant. In themselves they are often better than the national networks they interconnect.

— International packet services and a wide range of international private circuits and private networks are well developed in many parts of the world.
— International communications traffic and facilities have grown rapidly for some years, and show every sign of continuing to do so.

14 Customer premise equipment

The communications equipment on customers' premises (Customer Premise Equipment or CPE) comprises telephones, teleprinters, facsimile terminals, communicating PCs, Private Branch Exchanges and various related terminals and ancillaries like answering machines. There have been a number of important developments in the design of CPE in recent years. There have been similar developments in the constitutional basis on which it is provided in advanced countries. The provision and management of CPE has become a major undertaking in its own right, which can absorb substantial capital and technical and managerial resources. This chapter reviews the nature of CPE and discusses several managerial and policy issues to which it gives rise.

The equipment and its provision

A detailed review of the full range of modern CPE would be impossible within the scope of this book. The following paragraphs review the more important characteristics of modern CPE.

In the past CPE has been provided by telecommunications operators on rental. In Britain BT and its competitors also offer their customers the choice of purchasing or in suitable cases leasing their CPE. Purchase is becoming increasingly common.

Telephones and ancillary devices

So far as telephones are concerned the major technical development has been the introduction of high-impedance telephones. In traditional telephones the energy required to operate the bell was considerable. It had to be fed from the exchange through what is known as low impedance circuitry. Considerable energy was also required to make the microphone in the telephone work. The electrical situation was such that the telephone was in effect an integral part of the network and was normally permanently connected. It was possible to make arrangements for telephones to be connected through plugs, but the arrangements were clumsy and inflexible. Modern high-impedance telephones and the equivalent generation of other CPE are electrically much more independent of the network. They can be connected and disconnected at will. The policy significance of this is discussed later.

In addition, integrated circuits and modern memory techniques have been exploited to reduce the cost and increase the versatility of telephones. Facilities such as 'last-number-redial' are now standard on volume-production instruments. Many more advanced facilities like repertory dialling are available on more sophisticated instruments. De luxe telephones incorporate clocks and even calculators. The exterior design of telephones has also become varied and sophisticated.

Simple ancillary devices for association with telephones have been available in advanced countries for many years. Typical examples are answering and recording machines (which enable calls received in the customer's absence to be answered and recorded automatically); call-logging devices (which allow any desired combination of details about the length, destination and — in the more sophisticated designs — the cost of calls to be recorded automatically); and repertory diallers (which enable anything up to several hundred or even a thousand different dialling codes to be dialled automatically under the control of one, two or three buttons or some similar control device). There are many other varieties and combinations of ancillary device.

Plugs and sockets

The electrical separation between CPE and the network has made it possible to introduce really flexible arrangements for plugs and sockets. Terminals such as telephones, PCs and facsimile machines and many others can now be plugged and unplugged at will just like mains appliances. In the United States and Britain these arrangements have made it possible for telephones to be purchased directly from retailers who compete with the telecommunications operators. Customers who wish to have a telephone installed for the first time can order the exchange line and the socket from BT and purchase a telephone of their own choice either from BT or from a retailer.

These arrangements have proved attractive in Britain. Competition in supply of telephones is now well established. It is now usual for customers to buy and own their own instruments, although they can still rent them from BT if they wish.

Arrangements have also been made in Britain and in the United States for customers to do simple CPE wiring work for themselves, using approved kits available from retailers. This is called 'Do-it-yourself' (DIY).

In suitable circumstances such arrangements, which save the effort of the operating enterprise, could have attractions in developing countries.

Document terminals

There are several well-established telecommunications text communication services, each with its own terminals. Telex was discussed in Chapter 10. It uses teleprinters operating over the separate telex network. It is the best established text service. International telex has become particularly important. There are several reasons for this. Telex is the oldest standard text service. Its procedures and technical characteristics are substantially standard throughout the world; and it has developed in a way which eliminates the effect of clock time differences between zones.

In the last few years facsimile has developed very rapidly. Facsimile terminals connected to the PSTN can be used to send documents, pictures and other images anywhere in the world where there is a telephone line. It is estimated that there are now 2 million facsimile terminals in the world and the number is growing rapidly. Facsimile is established as a serious competitor to telex.

Both teleprinters and facsimile terminals have become more sophisticated in the last few years. Integrated circuits and micro-processors are exploited in modern designs. Storage facilities, which allow messages to be held over and sent when a circuit is available, and the equivalent of telephone facilities like repertory dialling are now common. PCs can now be adapted to function as telex or facsimile terminals (see below).

The general question of standards is discussed in Chapter 15. One point affecting text services should, however, be noted here. A standard promulgated by CCITT and designated 'X 400' relates to equipment providing facilities within the network to interlink the various text services. A service to the X 400 standard provides facilities for communication between telex, facsimile and all other document communication services, as well as mailbox facilities. BT has introduced a service of this kind. This is a very important development. It is strongly suggested that developing countries consider adopting the X 400 standard and providing services to it.

PCs with communications facilities

PCs with communications facilities now rank as communications terminals in their own right. Communications features (modems and the necessary software to control the communications process) are increasingly being incorporated in PCs and word-processors for use on analogue PABXs and exchange lines. 'PC Fax' boards are available which can be plugged into the back of PCs to equip them to communicate direct with facsimile terminals. If PCs are used on digital PABXs and exchange lines they do not of course need modems. Suitably equipped PCs can exchange documents and images with other computers or with central services like computer mailboxes over telephone or packet-switching lines; they can interwork with telex and facsimile terminals; or they can communicate with other PCs, or with mainframes to carry out computing processes or to access databases.

Terminal combinations

A number of manufacturers have experimented with combinations of modern terminals. One important development has been the so-called 'one-per-desk' terminal. This combines telephone, computing and data transmission facilities. Other permutations are now on the market, like combined telephones, facsimile terminals and answering machines.

Private Manual Branch Exchanges and Call-Connect Systems

The industry term for a private exchange installed on a customer's premises to connect his internal calls is a Private Branch Exchange (PBX). There are two kinds — Private Manual Branch Exchanges (PMBXs), which have human operators, and Private Automatic Branch Exchanges (PABXs).

PABXs range up to large installations serving several thousand extensions. They are normally used by organizations with more than, say, twenty extensions.

In past years smaller organizations often used PMBXs. In more recent years PMBXs have been largely superseded by key telephone systems (KTS). These are small networks of more complex telephones which can intercommunicate with one another without needing full PABX facilities or needing a human operator as a PMBX does.

In modern terminology small PBXs and KTS have both come to be referred to collectively as 'Call-Connect Systems' (CCS). There are many kinds of CCS. The simplest have no central unit at all, with all communication controlled simply by the electronics in the telephones themselves. On larger CCS installations there may be some kind of central unit, housing power supplies and the central control unit and to which the public telephone lines are connected.

Private Automatic Branch Exchanges

PABX technology has at least kept pace with the technology of public telephone exchanges. There are many SPC digital PABXs now on the market. These can all be arranged to provide a number of facilities in addition to basic communications. These include short-code, 'ring-when-free' and various other facilities associated directly with call connection, centralized dictation facilities, voice and text mailboxes and so on. They can also incorporate what amount to computer networking and main-frame computer-processing and storage capabilities. Suitably designed digital PABXs and their associated internal wiring can handle digital signals at speeds up to 64 kbit/s. The PABXs can be arranged to handle much higher speeds. Advanced network management systems for private networks analogous to those for public networks are now available on larger PABXs.

Many modern facilities like call diversion can be provided either from the public network or from PABXs. Indeed, particularly in the circum-stances of developing countries, these facilities may become available earlier on and be more easily provided from PABXs than from public exchanges. It may be cheaper and easier for the enterprise to encourage PABX suppliers to meet larger business customer needs in this way than from public exchanges.

Computer networking

There are very close similarities between mainframe and minicomputers and digital PABXs. Components, construction practices and operating principles are increasingly common to both. Indeed, the computing industry has introduced computers whose primary role was intended to be communications and which were for practical purposes PABXs. But the computing industry has also developed it own concepts for systems linking terminals and computers on customers' premises. 'Networking' arrangements, which enable computers and PCs to support one another (for example, by exchanging data files or by operating their processors in tandem for particular tasks — the technique called 'distributed pro-cessing'), are constantly gaining ground in computing.

Local area networks

A characteristic computing approach to internal networks is called the 'local area network' (LAN). A simple LAN* consists of a coaxial cable run round a customer's premises. Such a cable is very well suited to carrying

---

* The expression LAN is sometimes used with a wider meaning, to refer to any network which may be used to couple computers and PCs within a building or in a local vicinity. Some PABXs and CCS qualify as providing LANs in this wider sense. For WANS and MANS see Chapter 10.

digital pulses at speeds up to 1 mbit/s or even more. A series of microcomputers and one or more mini- or mainframe computers can be connected to the cable. Correctly designed and programmed computers can exchange messages on such cables with a minimum of extra equipment.

A LAN can carry speech as well as computer signals. There is no technical reason why LANs should not be used for speech instead of PABXs. But if the majority of the communications traffic in a premises is speech (as, today, it usually is) a PABX will usually be the most sensible choice.

Payphones

Public payphones are a feature of virtually all telecommunications operations throughout the world. In the early stages of telecommunications development public payphones are likely to be the first form of service available to the general public. BPOT provided public telephones on a subsidized basis from the 1930s to promote the spread of telephone service. As a result there are now a number of public payphones in places where they have virtually stopped being used because private telephones are now widespread. It is often difficult to persuade local residents to accept that a payphone should be withdrawn once it has been provided.

There are many designs of payphones available on the world market. Their coin-handling facilities can be adapted to most normal currencies, and they can be arranged to interwork with most modern exchange systems.

British experience is that difficult problems can arise with public payphones. Coin collections are difficult and expensive to arrange. The presence of substantial accumulations of coinage is likely to present a temptation to theft. For these reasons cardphones, operating either on special cards or on normal credit cards, have been developed and are now available on the market.

Public payphones are often installed in kiosks or similar shelters. These have to be cleaned, and they are liable to be abused in various ways. More generally, in countries where vandalism is a problem, public payphones and kiosks are likely to be an inviting target. Physically strengthened payphones and specially designed kiosks have been developed in Britain for these reasons. Mercury now competes with BT to provide public payphones at selected sites.

Customers can rent payphones for use on their premises, for example, in hotels and restaurants. A number of designs of customer payphones are available on the market.

Management considerations

CPE poses management and policy issues of its own in a number of respects.

From the point of view of management, the development, provision and maintenance of CPE is among the most difficult telecommunications work satisfactorily to organize and supervise.

Substantial resources are needed to develop a modern telephone, CCS or PABX. It is unlikely that any but a few developing countries will be able or will wish to develop their own. There are a very wide range of modern designs on the world market. Procurement can normally be by competition among several suppliers known to have suitable products. (Procurement practices are discussed in more detail in Chapter 12.)

CPE presents special problems of inventory (stock) control. Under modern conditions an enterprise which supplies CPE must carry a very wide range of stock, involving many hundreds or thousands of different items. In the case of telephones, for example, customers in advanced countries nowadays expect a choice of many different basic designs, each of which requires its own spare parts and must be stocked in a wide range of colours. More elaborate installations are likely to involve several different elements, like a central unit, outstations and a power supply. There are often several variants of such elements, with small differences between them. If the wrong variant is given to the installer it may not fit with the rest of the installation. A lot of time can be wasted unless exact type numbers and so on are closely monitored in the stores.

It is impossible accurately to forecast consumption of such an extensive and diverse range of stock. Computer-based control of stocks is essential. But even then stockouts and surpluses are inevitable. Reordering and surplus stock disposal procedures need to be very good if large working capital costs due to excessive inventory are to be kept within bounds and at the same time an acceptable service is to be given to customers.

Of its nature the installation and maintenance of CPE involves a considerable number of staff, working on their own or in twos or threes at most. The allocation and progress of installation work is usually controlled from central points (called in BPOT Installation Controls) by telephone or mobile radio. Such matters as the number of jobs done in a day and the routing of staff to and between jobs are governed hour-by-hour by these Controls. This has a big influence on efficiency and the Controls require close attention from management. But once the staff are on site the productiveness and quality of the work depends primarily on the conscientiousness of the individual workers. And quite often a job started, say, after lunch will be finished before the end of a working day; it is very much up to the individual what he then does. A Control may handle twenty or thirty installers or teams. But it is difficult for a supervisor to visit more than a few work sites in a day.

In BT the great majority of engineering staff on this work are responsible and competent people, with considerable pride of craft. They frequently forge excellent relations of their own with customers. But it is easier for a poor or unproductive worker to escape notice on this work than if he works in an exchange or a repeater station. Also there are inescapably opportunities for CPE staff to make money dishonestly, by doing off-the-record deals with customers to install what is often stolen apparatus. Such deals can be reduced by thorough vigilance and control, but this costs money and no system is watertight.

The need to support and oversee the operations of the CPE field force makes a number of other demands on the enterprise. Managers and technical support staff — including software specialists — who may be in short supply throughout the enterprise, will be needed in significant numbers if CPE work is to be properly run. The range of modern CPE is so wide that it makes special demands on training, which again must be met if the CPE work is to be done as it should be.

Competition in CPE and divestment of CPE work

The question of the admission of competition into CPE was referred to in Chapter 2. Historically in Britain as elsewhere, BPOT had a monopoly of provision and maintenance of CPE.Pressures from suppliers other than network operators, consumer pressures for greater freedom and choice of CPE, and the expectation that competition would force prices down, improve quality and lead to more vigorous exploitation of the technology led to the progressive admission of competition in CPE.

British experience is that the introduction of competition in this field has generally been successful. A much wider range of designs of terminals like telephones and answering machines is now available in the United Kingdom, compared with, say, ten years ago. British customers have access to the best of what is available on the world market. The BT share of the new instrument market in 1987–8 was below 80 per cent. The number of outlets from which CPE may be obtained has greatly increased. Price competition is strong. The range of large PABXs available to British customers has also notably increased and again they now have access to the best of what is available in the world.

The liberal trade policies of the British Government have played an important part in these developments. CPE suppliers can introduce terminals and PABXs which have been developed and manufactured abroad into Britain with confidence that they will be fairly treated as regards testing and approval (see below); and that they will not be disadvantaged or excluded by tariff and other barriers. There is no doubt that the British customer has benefited as a result.

The introduction of competition has its own problems, however. Some of the most important centre on the fact that users are entitled to expect that arrangements will be made for the testing and certification of CPE before it is put on the market. Testing may be confined simply to electrical safety and to an assurance that the device will not affect the working of the network once it is connected, for example, by transmitting signals which will interfere with network systems. But British experience is that customers will also expect some kind of certification that the device will do the job for which it is sold. And the enterprise itself has an interest in making sure that telephones and other terminals can receive incoming calls.

In these circumstances competition requires the establishment of testing and certification procedures which are technically valid and which are independent of the enterprise. It is important that competition is fair. If the enterprise continues as a competitor in CPE it will be inappropriate for it to be involved in the testing or approval of its competitors' devices. And the testing procedures also require the network operator to publish the technical specification for the conditions which the network presents and which the CPE will have to meet. He is not likely previously to have done this, and again considerable effort and expense may be involved. These requirements are clearly likely to pose particular problems for developing countries, who usually rely primarily on imported CPE and are unlikely to have their own testing institutions or to have published interface standards.

In Britain the test and approval functions have been separated. An independent approval board for telecommunications CPE (the British Approvals Board for Telecommunications — BABT) was established when competition was introduced. Tests are conducted by approved laboratories; approvals are then granted by BABT. Developing countries may consider using the services of testing institutions in other countries, even if they decide to set up their own approval bodies.

In the advanced countries the load on the approval institutions has proved to be very heavy in the early years following liberalization. Serious delays and backlogs of approvals can result. In the United States a procedure for terminals and CCSs was devised to meet this situation called 'self-certification'. The manufacturer of a new device is allowed to test it himself and to certify its conformance with published criteria. Legal safeguards are of course needed against irregular certification. Similar procedures have been considered in Britain. They could have real advantages for developing countries.

The processes involved in the approval of large PABXs for connection to the public network are very much more complex and demanding than those for terminals. They require special treatment. A large PABX is very similar to a public telephone exchange. Modern PABXs must be able to exchange a wide range of signals with their host public networks and they must conform to many other technical requirements. Thorough testing and evaluation of new designs is essential. Self-certification procedures of the kind applicable to terminals are not appropriate. On the other hand, it will be beyond the resources of many developing countries to establish their own PABX test houses. In these circumstances even developing countries who handle their own terminal testing may like to consider employing test houses abroad to test large PABXs for them. This cannot of course be done unless proper public network interface specifications are available.

Other problems with the introduction of competition can arise with maintenance procedures. In the vast majority of cases customers who encounter faults will report them in the first instance to the enterprise, whose maintenance staff will then have to localize the fault. It is not always easy to determine unambiguously whether the cause of a particular fault lies in the terminal or in the network. The problem is helped by plugs-and-sockets, since the telephone can be unplugged and substituted. But it has particular force in the case of large modern PABXs because of their complex relationship to the public network. Care is needed to make sure that arrangements involving CPE supplied by bodies other than the enterprise work smoothly.

There is another constitutional option besides competition. Telecommunications CPE is similar in many ways to devices like electric appliances, office equipment and so on which it is taken for granted customers will buy, lease or otherwise find and have maintained for themselves. Many of the problems of supplying, installing and maintaining consumer CPE are much more easily handled by small specialist firms like home electronics retailers or local electricians than they are by a nationally organized network operator. From the financial point of view it can be attractive for an enterprise which is short of investment to let customers find the capital for their own CPE. It was for this reason that prior to 1981 BPOT did not supply or install large PABXs at all. In principle, the same argument can apply even to straightforward telephones. It is on grounds such as these that at least one enterprise in an advanced country has relinquished CPE work altogether.

In most developing countries CPE has traditionally been provided and maintained by the network operator. The decision whether to admit competition or wholly to divest CPE is likely to be seen as a political one, which can only be taken at Government level. But it is important to bear in mind that the range and capability of the CPE on offer has a formative effect on growth in the use of the network and therefore on the finances of the enterprise. It is particularly important in influencing the development and exploitation of network services including VANS (see Chapter 15). Whether or not the enterprise retains a monopoly of CPE, therefore, it is important that its policies and those of Government foster imagination and enterprise in the provision of CPE.

Conclusions on management and policy

The analysis in preceding paragraphs suggests that an enterprise which is modernizing its telecommunications may have good grounds for reviewing its practice as regards CPE. It is important to remember that unless the enterprise withdraws wholly from a particular category of CPE it will still have to maintain the appropriate marketing effort, work-force, support structure and stocks to service whatever market share it sets out to keep, with all the attendant problems.

There may be particular arguments affecting the provision and maintenance of complex modern business CPE, like PABXs and data-networking equipment. This is usually costly and may require highly specialized skills to install and maintain. The operator's staff with such skills may be badly needed for work on the public network. In some circumstances it may be better to allow private firms to supply such apparatus even if the rest of CPE is to remain a network operator monopoly.

Special considerations of a different kind apply of course in remote or rural areas. In such cases it will often be impracticable for CPE to be supplied, installed or maintained by any one but the network operator, who will have staff in the locality anyway.

It is important that any decision which is reached about these matters is based on a thorough economic evaluation. It will often be appropriate for the evaluation to be carried out by the central planning unit of the enterprise, if one is established on the lines advocated in Chapter 3.

Summary

To sum up:

— There have been important technical and design advances in telephones, teleprinters and facsimile terminals in recent years.
— PCs are now communications terminals in their own right.
— PMBXs have been largely superseded by Key Telephone Systems (KTS). Small PBXs and KTS are now referred to collectively as Call Connect Systems.
— PABX technology has at least kept pace with public exchange technology. Large PABXs are now very intimately connected with their host public networks from the technical point of view.
— The computing industry has developed its own CPE networking techniques.
— CPE provision and maintenance work is different in kind from most telecommunications work, and poses its own management problems.
— Supplier and consumer pressures have led to the liberalization of CPE in a growing number of advanced countries.
— It may be appropriate for developing countries who are modernizing and expanding their telecommunications to consider withdrawing their enterprises from some or all sectors of CPE, or admitting competition. But liberalization or divestment of CPE work has its own problems and should be thoroughly gone into before anything is done.

15 Network services

In recent years a new category of telecommunications services has developed, provided from computers reached over the network. This merging of telecommunications and computing has great potential, especially for developing countries. In this chapter we look at these services and at some of the policy questions to which they give rise. We also consider visual services.

Value-added Network Services (VANS)

Computer-based services reached over the network add value to the use the customer gets from the network. They were first authorized formally in Britain in 1982. They are provided by competing operators including BT. For regulatory and constitutional purposes they are referred to in Britain and the United States as 'value-added network services' (VANS). They are sometimes called simply 'value-added services' (VAS). In Britain a particular class of VANS associated with data activities is called 'value-added data services' (VADS).

Some idea of the possible range and character of VANS is given in the Explanatory Note to the General Licence which authorized VANS in the United Kingdom in 1982. It encompasses: Automatic ticket reservation and issuing; conference calls; deferred transmission; long-term archiving; Mailbox; Multi-address routing; protocol conversion between incompatible computers and terminals; secure delivery services; speed and code conversion between incompatible terminals; store and retrieve message systems; telephone answering using voice retrieval systems; telesoftware storage and retrieval; text editing; user management packages, e.g. accounting, statistics, etc.; Viewdata (now called Videotex) services; Word-processor/facsimile interfacing. Most of these will be self-explanatory. Certain of them call for comment.

Database services

A database service offers its customers access to information stored on a central computer, and the facilities of that computer to locate what they want. A wide range of such services is now used by people like librarians and information officers. Specialist database services are growing rapidly in Britain.

Deferred transmission services

These are similar in principle to the telex store-and-forward facilities described in Chapter 10. They can cater for any kind of traffic, including voice messages and facsimile, as well as for telex and teletex and PC communications.

Protocol and code conversion services: the standards problem

These services and the requirement for them give rise to some important issues. These are broader than the subject of this chapter, but it is convenient to consider them together here.

For various reasons firms in the computing industry have always used their own proprietary standards for hardware and software. 'Communications protocols' is the technical name for a set of detailed rules for communications governing things like the content and meaning of signals, layout of documents and so on. Various protocols of this kind are embodied in proprietary standards. While computers remained stand-alone devices, or were supported only by limited networks of their own servicing only proprietary terminals, this could create local problems for people coupling computers and peripherals. But otherwise it did not matter.

But more recently serious complications have begun to arise. Situations are now common in advanced countries in which users or service operators need to be able to link terminals computers and peripherals produced by one firm with those produced by others at any distance up to intercontinental. Incompatible terminal and computer standards make this a complicated and expensive business. Problems of incompatibility are therefore becoming serious and are clearly liable to affect developing countries. There are several ways in which they might be resolved.

The ideal situation would be one in which all manufacturers produced hardware and software to a single set of interface standards. This would make it possible for users to interconnect any design or type of equipment with a guarantee that it would work properly. From the user's point of view this would be a major advance. The situation would be analogous to that which has always existed in public telecommunications networks. It is fundamental to the working of telecommunications that each exchange should be designed to work properly with all the other exchanges with which it is in communication and with the transmission links between them. Telecommunications manufacturers and operators expect to design their equipment to do this. The traditional approach of telecommunications operators is to settle interworking standards by collective discussion through the CCITT or equivalent bodies.

A major programme has been mounted by the International Standards Organization (ISO) in cooperation with the CCITT to develop and agree a single world-wide family of computing interface standards called Open Systems Interconnect (OSI). This work involves computing, telecommunications and office machinery interests of all kinds.

OSI has its own problems. It is a much more difficult task to define a set of standards which will ensure that non-voice terminals can 'talk' to one another in terms of such details as the precise effect of keys, the layout of documents and technical computing procedures, than it is to define standards just to enable different designs of telephone exchange to circulate speech between them. Nevertheless OSI is steadily gaining momentum, especially in Europe.

In the meantime the computing industry is continuing to release hardware and software based on proprietary standards. As a result there are so many incompatible devices in customers' possession and so many new ones still being bought that there is a strong incentive to devise more *ad hoc* solutions. The simplest is to develop special interfacing hardware and/or software to carry out all the operations needed to convert the output of one particular design of computer or PC into a form another one can accept. An immense variety of devices of this kind has been developed and put on the market. The range grows all the time.

Individual PCs and other terminals which need conversion hardware facilities to intercommunicate may only need to do so intermittently. A considerable amount of money may be saved by arranging for them to use a common pool of interfacing hardware. In a larger network — say, in substantial office premises — it may often be sensible to locate this pool at the PABX or at some other central point in the network, such as a mini- or mainframe computer. Where terminals and computers are required to communicate over public networks or over longer-distance private networks, the conversion facilities can be centrally located in those external networks. Where such facilities are provided by operators as a service for third parties, of course, they constitute one form of protocol conversion VANS.

An important and growing telecommunications application for such interfacing techniques is between public document communications services. The promulgation of the X 400 standard within the OSI matrix has been an important step forward. This is an interfacing standard based on the principles of OSI. A communications node designed to this standard will have interfaces which will enable it to circulate traffic between all the text services — telex, teletex, facsimile, electronic mailboxes and so on.

All such approaches involving special interfacing hardware and software are, however, essentially a compromise. The extra hardware and software cost money, add complexity and take time and effort to arrange. There is another approach. It should be possible to use the reasoning powers of artificial intelligence in the intelligent networks of the future (the UICN — see Chapter 10) to make them adapt themselves to interface with equipment designed to a variety of original standards.

These problems of standards and interworking have been attracting attention in the world computing and telecommunications communities for some time. Concern is growing about their impact on customers and on the spread of information technology. It is difficult to predict which of the various approaches will predominate in the long run. But from the point of view of telecommunications enterprises and users, the preferable approaches are clearly OSI and self-adapting networks. There is no reason why the two should not exist alongside one another.

Developing countries acting either singly or together should be able to exert considerable influence on these problems as suppliers turn increasingly to them for markets.

Videotex services and telematics

Videotex services are a very important group of VANS.

Prestel

Prestel was the first Videotex service. It was devised in Britain. In the earliest form of the service the user needed only an adaptor which he could insert between his aerial and an ordinary television set. It was connected to an ordinary telephone line. When the user pressed the appropriate keys on the control unit, a call was made automatically over the PSTN to a central computer. By pressing other keys the user could call up any one of several hundred thousand pages of information. Dedicated Prestel terminals, with keyboards, a screen and communications facilities have now been developed for frequent users. Alternatively any PC can be used as a Prestel terminal. The user pays a standing charge for connection to Prestel, a charge for some of the pages he calls up depending on who provides them and a normal telephone call charge.

In Britain Prestel is run by a division of BT. The information on its computers is provided by other interests. A number of organizations have closed user groups (CUGs) on Prestel. Such groups have access to information on the central computers reserved for their private use, and not available to users generally. Many of the members of these closed groups have dedicated terminals. Over 95 per cent of the members of the Association of British Travel Agents have terminals of this kind which they use to call up information about travel and hotel bookings fed into central computers by tour operators and similar bodies.

Prestel was originally seen as a service for the business community. Because it was such a simple service to provide and because it was conceived round the telephone and the domestic television set, however, it was subsequently felt that it could be popular in a much wider market. Plans were drawn up to market the service to residential as well as business customers. But in the event it has developed primarily as a business service. By mid-1988 Prestel had a total of 80,000 customers in Britain. It also has subscribers in over thirty other countries, who reach it over the international telephone network.

Teletel in France

In a number of countries outside Britain services which provide mass users with access to central computers over public switched networks in this way are called 'telematics'. For reasons discussed more fully below a different approach to telematics from that of Prestel was adopted in France. After various experiments and field trials the Direction Générale des PTT introduced a substantive service in 1982 called Tèlètel. When it started the service provided access to business database services over the French domestic packet-switching network (Transpac). This was significant not least because the Transpac network charged for calls on a basis independent of distance, in the same way as the BT PSS (Chapter 6). An electronic directory service for all customers began operations in Rennes, a major provincial centre, in early 1983.

The basis of the Tèlètel system is that the terminals (Minitels) are distributed free to subscribers, whereas Prestel customers must pay for theirs. There seems little doubt that this is among the reasons why the Tèlètel service, unlike Prestel, has expanded in a dramatic way. By the end of 1987 almost 3 million Minitels had been distributed to subscribers and the service is extensively used by both residential and business customers for a variety of database and messaging applications. The relative growth of Prestel, Tèlètel and Bildschirmtext (the West German service which uses a similar format to Prestel) is compared in Table 15.1.

Sector applications: financial services and EFTPOS

So far we have been considering examples of VANS from the list in the original British General Licence. That licence is framed in terms of services rather than sectors. One business sector is already a major user of VANS, and indeed of modern telecommunications generally. This is financial services. A wide variety of financial services is already operating, and rapid progress continues with the exploitation of VANS in this sector.

Table 15.1 Relative growth of Teletel, Prestel and Bildschirmtext

	1982	1983	1984	1985	1986	1987
France: Tèlètel						
Subscribers (end of year)	—	120,000	530,000	1,300,000	2,200,500	2,791,000*
Services (end of year)	—	—	840	2,070	4,150	5,662*
Hours of connection (m.)†	—	—	—	1.0	2.5	4.7*
Britain: Prestel						
Subscribers	19,850	38,000	48,000	63,000	70,000	76,000
Information providers	1,003	1,356	1,365	—	—	1,252
Number of frames	41,050	277,100	330,000	320,000	300,000	310,000
Frame calls per week (m.)	—	—	3.4	7.6	9.1	9.1
FR Germany:						
Bildschirmtext Subscribers	—	10,555	21,319	38,894	58,365	83,633
Information providers‡	—	2,740	3,099	4,043	3,528	3,416
Remote databases	—	0	37	1,511	218	248
Number of frames	—	378,000	521,783	762,673	589,330	610,704
Number of calls (m.)†	—	0.1	0.3	0.5	1.1	1.9

*Data refer to June 1987.
† Per month.
‡ Sub-information providers are included.
Source: Mayntz and Schneider (1988).

Telecommunications has had a particularly marked impact on the international money market. Most international currency dealings in advanced countries are now conducted instantaneously over telecommunications links. Great ingenuity has been devoted to the development of specialized equipment and network facilities to match the needs of this extremely demanding market. The principal stock markets and commodity exchanges of the world also now depend for their operation largely on telecommunications and computing facilities used for the distribution and retrieval of information — for example, share prices — and for the actual conduct of transactions.

In Britain, the major high street banks have used large mainframe computers for their central book-keeping and accounting networks linking branches and head offices for many years. Most internal banking transactions are now carried out on these networks. It is relatively simple for two banks which have internal networks of this kind to arrange to interlink them, so that inter-bank transactions may be carried out in the same way. This is now well-established practice between financial institutions of many kinds. Movement of funds by these means is called Electronic Funds Transfer (EFT).

Cash-dispensing terminals (Automatic Teller machines or ATMs) are now installed outside many high-street bank branches in the United Kingdom. When the appropriate card is inserted and an identity code has been keyed, customers are able to draw up to, say, £100 in banknotes, enquire about their balance or ask for statements at any time of the day or night. These terminals are linked to the banks' central computers over their private circuit networks.

Many credit cards now have a magnetic strip. A simple application of this is to allow the card concerned to be validated by a terminal on a retailer's counter. The card is inserted into a specially designed terminal which communicates with the credit-card company computer over the public network. The strip is read and signals identifying the card are sent to the computer. If the card is valid, the computer returns signals giving clearance.

Similar arrangements can be used to notify the computer instantaneously of the credit-card transaction, and to debit the customer's account. In this way the cumbersome process of completing a paper record and sending it to the credit-card company which subsequently sends a bill to the customer can be eliminated. This technique is called electronic funds transfer at point of sale (EFTPOS).

Technology for EFTPOS has been available since the late 1970s. In Britain operational EFTPOS schemes are only now beginning to be established. Among the reasons for the time it has taken to introduce EFTPOS are, first, the need for guarantees of absolute reliability on any transaction involving the transfer of real funds; and, secondly, the complex organizational and commercial problems which had to be overcome in introducing a widespread scheme involving retailers, banks and customers.

It was natural for the financial services sector to exploit the new facilities. Information is the essence of finance and accounting. Banks and similar institutions had used mainframe computers for routine book-keeping and mass clerical processes for many years. Telecommunications and computing equipment have a high capital cost, but properly used they can save running costs and greatly enhance productiveness. The financial sector is well placed to take advantage of these characteristics, and well attuned to them.

The exploitation of the new capabilities has been stimulated and accelerated by the competitive nature of the business. The effect on the banks' competitiveness has turned out to be such that as soon as one major bank around the world began seriously to use information technology, it could only be a matter of time before the others were obliged to do so. The effect has been greatly to speed up, and indeed to alter, the character of financial dealing of every kind. The financial services and information technology industries have become deeply interdependent.

The British insurance industry is now beginning to exploit VANS. Services are available which provide insurance and investment brokers with access from their own terminals to insurance and investment information on centrally organized mainframe computers.

Both technology and banking techniques are now available to support the rapid spread of electronic funds transfer to other sectors. These are likely to be sectors which include strongly competing operators with considerable resources and open to international influences, like major retailing chains; and retailers with special cash-handling problems like petrol (gas) stations.

Policy issues: the mass use of telematics

These applications in advanced countries are confined to relatively limited and specialized sectors with natural affinities to improved ways of circulating and handling information. Developing countries have special reasons to be interested in the mass application of VANS and of one group of these services in particular. This is the group available to the public at large over the PSTN or the IDN. These services are called 'telematics', or 'telematique' in French.

In *The Telematic Society* published in 1980 (Martin, 1981) the distinguished author James Martin described the nature and potential of telematics. In Chapter 23 he drew particular attention to their potential for developing countries, and to the social and political issues raised.

These questions are important to developing countries. Properly used, telematics would make possible a number of important advances. Outlying communities could benefit greatly from cheap and simple access to medical, agricultural, educational, transport and other information stored on central computers. Growing business enterprises in manufacture, mineral extraction, agriculture and many other fields stand to benefit considerably if their accounting, order-taking and general business correspondence are carried out by telematic services. In difficult terrain or where large distances are involved facilities for instant communication and processing of written messages, orders and their confirmation and many other documents will have special value. In principle, once telecommunications service is available all these services can readily be provided by the properly organized use of telematics.

In 1978 an important study of the development of telematics was published in France called *L'informisation de la Société* by Simon Nora and Alain Minc. An English-language edition was published by MIT Press in 1980 (Nora and Minc, 1980). The introduction to the latter includes the following:

[In June 1979] there are before the Congress three Bills to modify the Communications Act of 1939. If passed these would open up the telecommunications field to more competition and give the market more voice in shaping the development of 'communications' (telematics). To that extent the direction of the United States is opposed to that of France, where the effort — if the Nora/Minc report is implemented — will be to give the Government a more active role in the introduction of telematique (telematics).

In the event developments in the United States and Britain did open telecommunications to more competition and to the working of the market. So far there has been less change in the regime in France (although private sector provision of customer apparatus has always been permitted). It is interesting to compare developments in telematics and Videotex in the two countries since the books referred to above were published.

As we have noted, in Britain VANS have been very vigorously exploited by or on behalf of very large companies and state organizations and in specialist sectors like finance and publishing. Many big organizations are now dependent on VANS for the conduct of their operations. In the mass market there has been a extensive take-up of 'stand-alone' PCs and word-processors (see Chapter 14). But genuine telematic applications for the mass market like Prestel have been slow to take off.

On the other hand, the success of the comparable French Tèlètel service in the mass market has already been noted. The arrangements for the Teletel service reflect the thinking in the Nora/Minc report.

Conclusions on policy for telematics

Originally, the market for telematic services was seen as centring on the capability they offer for rapid access to large quantities of information by mass customers. But the accumulating experience in Britain and France suggests that in the advanced countries the true future role of telematics may be primarily as transaction services used either with 'dumb' terminals or with PCs, depending on the penetration of the latter in the country concerned.

The nature of this transaction market for mass telematics in advanced countries is gradually becoming clearer. There seems to be little doubt that PC-based text services with or without mailboxes ('electronic mail' — see Chapter 14) will gradually gain ground, even compared with facsimile. Banking transaction services from home or office providing facilities for customers to check balances and transfer funds and travel enquiry services of one kind or another all seem likely to develop in a solid way. As transport congestion increases the demand for transaction shopping services and similar applications could also expand in a major way.

The situation in developing countries is different in important respects. In many of them the general population is far from affluent, and unfamiliar with information technology even in a simple form. Personal computers are likely to be scarce because they are relatively expensive and complex to use. A fully capable PC with 30 or more megabytes of storage costs roughly ten times more than a 'dumb' telematics terminal; and it needs considerable expertise to use it properly. In the majority of developing countries a central computer reached over the PSTN from 'dumb' terminals with simple operating procedures should be a much more rapid, cost-effective and accurately targetted way of bringing information technology to the mass market than PCs.

The French model is interesting from the point of view of developing countries for another reason. In a country with a less affluent population the most effective way of exploiting telematics may be through state channels, with distribution of terminals free or for a low charge.

Only the developing countries can decide how they approach these matters. But in the circumstances of many of them the Tèlètel approach clearly deserves careful thought. Once telecommunications service is available in a given area it would be possible to provide Telematic terminals at state expense (or at the expense of an aid agency) on the basis of, say, one per enterprise and one per outlying community. These terminals could be used to reach central computers providing comprehensive database, processing and transaction facilities designed specifically to meet the needs of the enterprises and communities concerned. If and when mass demand developed, more terminals or if appropriate PCs could be coupled into the network with no difficulty.

The key to such an approach will be a centrally organized initiative. The telecommunications enterprise is likely to be well placed to mount such an initiative. It combines access to state funds, the power to define communications and other technical standards, the ability to adapt features of the telecommunications network like charging patterns to suit telematics, and a network of skilled people across the country to install terminals and demonstrate how they work.

British and French experience has shown that this is a complicated field, in which developing countries would be well advised to take expert advice before they act. A particular set of problems are likely to arise in arranging satisfactory telematics services and in managing databases and keeping them up-to-date. And it would be easy for any scheme, whether on the Videotex or Tèlètel model, to break down because the charges were too high, the procedures were too complex, or because the hardware or software was unreliable. Nevertheless the potential of mass telematics is great and their application deserves thorough consideration.

Visual services

Although they are not necessarily VANS, no consideration of network services would be complete without a look at visual services — that is, communications services for moving pictures. At present these services present something of an enigma. For example, techniques have been available for some time which would permit conferences to be held at a distance over television channels and would enable callers making telephone calls to see one another. But demand for video-conferencing facilities has been slow to develop, and no operator in the world has so far introduced a public switched vision phone service.

Video conferencing

BT's main transmission network has been used for a long time to convey BBC and IBA programmes from studio to transmitter throughout the country. Country-wide television would not have been possible without the creation of a broadband analogue transmission network in the late 1950s and early 1960s.

Techniques were developed some years ago to enable two or more groups of people assembled in different centres to be linked by vision and speech circuits, so that a conference could be conducted between them. BT introduced such a video-conferencing service, called Confravision, in the 1970s. It used specially equipped studios located in BT buildings in London and a number of major provincial cities. The pictures were sent over fully equipped standby channels on the main television programme network. Users had to come to the studios.

Some companies have made limited use of the service over the years. The most extensive user has been BT itself. Large-scale use of the services has, however, failed to develop. Among the reasons are no doubt the high cost of the service (related to the very expensive transmission and studio facilities involved) and the fact that users had to travel to BT buildings in order to use it. Mobile studios have since been introduced.

My own experience suggests that in addition there is a factor related to users' sense of the limitations of television as a medium for visual contact. In BT in the 1970s Confravision facilities were extensively used for budgetary and other similar discussion between headquarters and field operating units. But it rapidly became clear that the participants felt they were only a substitute for personal contact. This had to do partly with the limitations of the particular camera and screen equipment employed at that time. These made it difficult to see facial expression and other visual reactions in the same detail as that taken for granted in across-the-table conversation. But it also undoubtedly had to do with a more elusive lack of sense of personal contact, and with the desire for casual interchange of a kind not possible in the structured environment of the service.

Much attention has been given in recent years to the possibility of compressing or processing television signals to reduce transmission costs. Compression techniques have now been developed which will enable pictures fully acceptable for business conferences to be transmitted over digital links at 2 mbit/s (where conventional television requires 68 or 140 mbit/s). Systems are available that will transmit acceptable pictures at rates down to 256 kbit/s. Vision terminals are now being developed that will transmit reasonably acceptable moving pictures at 48 kbit/s, so that the signal will fit alongside speech at 16 kbit/s in a single 64 kbit/s channel (see Chapter 9).

An alternative approach starts from the fact that digital transmission link costs are dropping sharply. With the progressive reduction in the costs of 2 mbit/s private circuits, it could well be cheaper to install the relatively cheap equipment required to compress a television signal only to 2 mbit/s and to rent a 2 mbit/s channel, than to install the relatively expensive equipment needed to compress it to 256 or 48 kbit/s and to rent the corresponding channel.

Satellite circuits are well suited to video conferencing. For example, companies can rent or purchase relatively simple camera and screen equipment which can be installed in normal rooms, and then make arrangements with one or other of the satellite carriers for dishes to be installed on their own premises. These dishes can communicate directly with similarly equipped premises at the distant end. We may expect continuing rapid growth in such arrangements, competing with more conventional video conferencing over terrestrial lines. Such arrangements may of course suit well the circumstances of developing countries.

Overall, cheap and flexible video-conferencing is now possible. There is no doubt about its value in certain defined applications in business and administration and education.

Vision phone

From the technical point of view it has been possible for some years to arrange full switched vision-phone service — that is to say, service which will allow the parties to a normal telephone conversation to see one another.

No operator in the world has so far introduced a regular public switched service of this kind, however. The major technical obstacle has been the requirement to provide large-scale facilities to switch the high bandwidth or bit-rate video channels involved. In principle digital switching systems can do this. But even with digital technology the capital cost of the combined switching and transmission facilities needed to provide even a limited public switched service has been seen as prohibitive.

Other reasons for the lack of implementation of vision-phone systems certainly include the high cost of camera and display equipment and its relative limitations. There is now intense competition at world level to develop cheap and physically small 16 kbit/s speech + 48 kbit/s vision terminals of the kind referred to above. These would be ideally suited to provide switched vision phone over the new 64 kbit/s networks, if costs can be sufficiently reduced. This is a promising area of development, but it is probably one where developing countries would be best advised to wait and see what happens, rather than making plans at present.

Telecontrol, Telemetering and Alarms

Three other long-established network facilities should also be mentioned here. Telecontrol facilities allow devices such as sluice-gates on a reservoir, radio and radar installations, and many others of similar kind to be operated from remote control rooms. Telemetering facilities allow readings from measuring devices at such locations to be read similarly from central control rooms. Automatic burglar alarms and security devices, connected to police stations by private circuits or capable of making automatic calls over the public network, are also well established in advanced countries.

Summary

To sum up:

— Value Added Network Services (VANS) have developed as an important application of the combination of telecommunications and computing.
— Serious problems have developed world-wide in connection with interface standards. They are impeding exploitation of the technology. Developing countries are well placed to press for their resolution.
— The financial services sector has dramatically demonstrated the potential of VANS.
— Telematics are potentially very important to developing countries. The approach adopted in France deserves close study.
— Until recently visual services have been slow to take off. But technical developments may change this before long.

Part III Conclusions

16 Conclusions

This Chapter reviews the principal conclusions to be drawn from earlier chapters. The theme of the book is that telecommunications should be treated as a modern business undertaking with a key role in modern society and economic activity. As such, it deserves its due share of capital and human resources. Given this and if it is properly managed, telecommunications can make a major contribution to a country's growth and advancement. But to make such a contribution the telecommunications enterprise needs to be run as an independent high-technology business. In the present state of the technology it is also likely to be a practical monopoly across most of its operations. It needs to be managed in a way which takes account of the special problems this poses. The purpose of earlier chapters has been to analyse British experience to assist countries who are setting out to do all this.

Principal conclusions

Chapter 2: Constitutional issues

The question of state or private ownership of telecommunications operators is briefly discussed. The strong reasons why telecommunications should be run as a modern business enterprise separate from the central machinery of Government and from posts, tourism and other activities are reviewed. British experience with competition in telecommunications is examined. It is concluded that in developing countries the main place for competition is probably in provision, installation and maintenance of customer premise equipment and value-added network services. The special problems of managing the rest of the business as a monopoly public utility are examined.

Chapter 3: Organization

The questions that arise in organizing a telecommunications enterprise are examined in detail. They have much in common with those that arise in business generally. But British experience has identified certain issues of particular importance in telecommunications. The first is the organization of field operations and their relationship to the Central Headquarters. So far as can be judged from British experience, organizations based on a substantial degree of devolution of authority to field units seem likely to be best for developing countries other than those with particularly limited territories or which are at a relatively low level of development.

The other issues have to do with the number of tiers of field organization, the management of the network, the separation of the running of the network from the development and marketing of applications, the choice of functional or 'product' group organizations, channels of communication with customers and the management of new products and services. All these matters are discussed. Organization of international services, buildings work and number information services are also considered.

Chapter 4: Accounting and control systems

and

Chapter 5: Plans and planning

Proper accounting and statistical systems, organized mechanisms of answerability and orderly planning procedures are all indispensable if an enterprise is to be properly run. Approaches to these matters suggested by BT experience are examined in detail. Mechanisms for target setting and answerability and the role of one year and five year plans and their compilation are considered in detail.

Chapter 6: Charging and pricing

Arrangements for charging and for the pricing of telecommunications depend to an extent on local requirements and circumstances. The basics are, however, examined and certain principles based on British experience are proposed. Developing countries are in a good position to review and innovate in charging arrangements, especially for modern non-voice services (text, data and so on). The questions that would arise in such a review are discussed in detail.

Chapter 7: Priorities and internally generated capital

A series of priorities are suggested for action by countries who wish to improve their telecommunications. Much can be done by proper management of existing plant. The capital needs of an enterprise should be assessed on a year-by-year basis by overall planning, rather than on a project-by-project basis. The proportion of capital requirements met from internally generated funds can be maximized by skilful management. Techniques for doing this are examined. But internally generated funds are likely to fall well short of the full need. At bottom the issues of external capital funding are political. They need to be addressed in the light of the conclusions of the Maitland Commission.

Chapter 8: Staff and staffing

It is stressed that, even though it is a capital-intensive high-technology business, success or failure in telecommunications depends in the end on people. British techniques for determining requirements for recruiting, training and organizing engineering, operating, office, computing and management staff are described. BT approaches to the improvement of staff productiveness (productivity) are set out. Broad philosophies of staff management and industrial relations are briefly discussed.

Chapter 9: The technology in outline

Telecommunications and computing technology can readily be understood by non-technical people in sufficient depth to allow them to take policy and management decisions. The chapter contains an outline of the essentials designed for such readers. The principles of computing and telecommunications hardware and software, the digital and analogue principles, and modern developments like supercomputers and optical-fibre cables are all dealt with.

Chapter 10: Networks

The nature of the various telecommunications networks — PSTN, telex, packet-switching, private networks, LANs, MANs, combined telecommunications and cable television networks and so on is discussed. In most countries the principal communications networks, including the PSTN, are changing radically as they evolve from analogue to digital technology. Developing countries would probably do best to concentrate on the basic conversion of their networks to integrated digital operation — the IDN stage — and to wait until market prospects are clearer before they consider offering full Integrated Services Digital Network facilities to their customers.

Chapter 11: The role of radio

Terrestrial radio customer distribution systems, microwave main transmission systems, satellites, cordless, 'Telepoint', personal communicators and mobile systems are all described. The significance of spectrum constraints, cellular techniques and millimetric frequencies are explained. The broader significance of radio as a distribution medium in high-density areas and its advantages over cable are considered. Radio systems have an important part to play in telecommunications every-where nowadays, not least in developing countries.

Chapter 12: Planning, procurement and management of network plant

The techniques of forecasting load on, dimensioning and laying out networks of all kinds including modern intelligent networks are dis-cussed. Surpluses and shortages of network plant have major implica-tions for service and for the economics of the enterprise. Networks need to be actively managed to minimize them.

British procurement practices for network plant are outlined. These practices require close senior management attention. They need to be demonstrably fair and subject to audit. Developing countries need sufficient technical expertise to be able to assess what they are offered and to carry out their own quality checks.

Chapter 13: International telecommunications and the ITU

The constitution and functioning of the ITU and the Centre for Telecom-munications Development are outlined. The way in which the ITU and the Centre focus help to the developing countries from the international telecommunications community is described.

Good international links and good supporting inland networks serving business and industry are vital to the economies of most developing countries. The character and functioning of international telecommunica-tions plant are described. The technical problems of international communication have been solved. International traffic and services have grown rapidly for some years and are likely to continue to do so. It is important that the capacity and quality of plant keeps pace.

Chapter 14: Customer Premise Equipment (CPE)

The important technical and policy advances in Customer Premise Equipment in recent years are outlined. Plug-and-socket arrangements for telephones are well established in Britain. PCs are now communica-tions terminals in their own right. Text services are evolving rapidly. PABXs have at least kept pace with public exchanges. CPE work is different from most telecommunications activity and poses its own management and policy problems, which are discussed. The reasons why developing countries should consider admitting competition in CPE, or even the complete withdrawal of network enterprises from CPE work are explained.

Chapter 15: Network services (VANS)

The essentials of Value-Added Network Services (VANS) are discussed. They have developed as an important application of the combination of telecommunications and computing. They are potentially of great importance for developing countries. Telematic services are particularly relevant. The approach of France to these matters (the Tèlètel or Minitel system) is outlined and contrasted with Prestel and Bildschirmtext in West Germany. It is suggested that the French approach deserves close attention by developing countries.

Telecommunications is not only an indispensable element of a modern society and a thriving economic system. It is also one of the most rewarding public utility disciplines. The community sets a high value on cheap, efficient communication between Government and the people; between businesses and their correspondents and customers; and among people and their families and friends.

Appendix 1

Extract from the report of the Maitland commission

CHAPTER 10

CONCLUSIONS AND SUMMARY OF RECOMMENDATIONS

1   The considerations on which we have based the response to our Mandate have been the subject of previous chapters. In this chapter we set out the conclusions we have drawn.

2   The telecommunications situation across the world has certain notable characteristics. Advanced industrialised societies have virtually comprehensive services. In developing countries, services are mainly concentrated in urban centres. Continuing technological advances offer ever increasing efficiency, reliability, and lower unit costs. The level of investment in telecommunications in developing countries is generally low. With certain notable exceptions, telecommunications services in many developing countries are poor or indifferent. In many remote areas there is no service at all.

3   Given the vital role telecommunications play not only in such obvious fields as emergency, health and other social services, administration and commerce, but also in stimulating economic growth and enhancing the quality of life, creating effective networks world wide will bring immense benefits. An increase in international traffic will generate funds which could be devoted to the further improvement and development of telecommunications services. The increased flow of trade and information will contribute to better international relationships. The process of creating effective networks world wide will provide new markets for the high technology and other industries, some of which are already suffering the effects of surplus productive capacity. The interest industrialised and developing countries share in the world-wide development of telecommunications is as great as in the exploitation of new sources of energy. And yet it is far less appreciated.

4   We look to governments of industrialised and developing countries alike to give fuller recognition to this common interest and to join their efforts to redress the present imbalance in the distribution of telecommunications which the entire international community should deplore.
5   We have identified several key elements in the joint effort for which we appeal.

— First, governments and development assistance agencies must give a higher priority than hitherto to investment in telecommunications.
— Secondly, existing networks in developing countries should be made more effective, with commercial viability the objective, and should become progressively self-reliant. The benefits of the new technologies should be exploited to the full to the extent that these are appropriate and adaptable to the countries' requirements.
— Thirdly, financing arrangements must take account of the scarcity of foreign exchange in many developing countries.
— Fourthly, the ITU should play a more effective role.

6   Our recommendations reflect this analysis of the problem and are aimed at stimulating the actions we consider essential if progress is to be made in creating effective telecommunications networks world wide.
7   First, to ensure that telecommunications are given the priority we believe they deserve, WE RECOMMEND that

a)   developing countries review their development plans to ensure that sufficient priority is given to investment in telecommunications (Chapter 9, paragraph 9);
b)   developing countries make appropriate provision or telecommunications in all projects for economic or social advance and include in their submissions a checklist showing that such provision is being made (Chapter 9, paragraph 10);
c)   countries and international agencies with development assistance programmes ensure that specific provision is made for appropriate telecommunications facilities in development assistance projects (Chapter 9, paragraph 21):
d)   contributors to and beneficiaries of the UNDP reconsider the importance they attach to the telecommunications sector, and provide appropriate resources for its growth (Chapter 3, paragraph 12).

In addition to these specific recommendations, WE APPEAL to the governments participating in the next Economic Summit to give encouragement to practical measures to improve and expand telecommunications (Chapter 9, paragraph 23).
8   Secondly, to make existing networks in developing countries more effective and progressively self-reliant and to exploit the benefits of the new technologies, WE RECOMMEND that

a) telecommunications operators in developing countries review their training plans (Chapter 6, paragraph 18);

b) developing countries use the resources available through the IPDC (Chapter 6, paragraph 19);

c) industrialised countries organise seminars to improve the qualifications of experts from developing countries (Chapter 6, paragraph 20);

d) the ITU supplement the catalogue of training opportunities with information about training opportunities in the private sector (Chapter 6, paragraph 22);

e) operators and manufacturers consider how they can enhance the training opportunities they offer to developing countries (Chapter 6, paragraph 23);

f) the major regional and sub-regional political and economic organisations consider as soon as possible how best research and development institutes might be established (Chapter 7, paragraph 15);

g) the research and development institutes proposed be developed as a source of higher technological, supervisory and managerial training and as coordinating agencies for external training opportunities (Chapter 6, paragraph 21);

h) developing countries consider pooling their purchases of appropriate equipment including terminals and components (Chapter 5, paragraph 24);

i) when purchasing equipment, developing countries ensure that the contract includes commitments on the supply of spare parts, training, comissioning, post-installation and maintenance (Chapter 5, paragraph 25);

j) manufacturers and operators be encouraged to develop systems which will enable the needs of the more remote areas of developing countries to be met at lower cost (Chapter 4, paragraph 30);

k) the ITU, in conjunction with manufacturers of telecommunications equipment and components, consider compiling a comprehensive catalogue of telecommunications suppliers and systems currently in use (Chapter 4, paragraph 33);

l) developing countries review the possibilities for local or regional manufacture (Chapter 7, paragraph 22);

m) manufacturers in industrialised countries consider the scope for cooperation with developing countries in local or regional manufacture (Chapter 7, paragraph 23).

9   As an immediate step to improve the present arrangements for assisting developing countries WE RECOMMEND that;

a Centre for Telecommunications Development, with its three components of a Development Policy Unit, a Telecommunications Development Service and an Operations Support Group, be established by the Administrative Council of the ITU during 1985 (Chapter 8, paragraph 4).

WE INVITE the Secretary-General of the ITU to carry out the necessary consultations so that the Centre can be established as soon as possible in the course of 1985 (Chapter 8, paragraph 15).

10 Thirdly, to finance the development of telecommunications WE RECOMMEND that

a) countries and international agencies with development assistance programmes give higher priority to telecommunications (Chapter 9, paragraph 20);

b) those who provide international satellite systems study urgently the feasibility of establishing funds to finance earth segment and terrestrial facilities in developing countries (Chapter 9, paragraph 22);

c) industrialised countries extend export/import financing and insurance cover to suppliers of telecommunications equipment (Chapter 9, paragraph 25);

d) the IBRD consider including telecommunications in its proposal for multilateral guarantees against non-commercial risks (Chapter 9, paragraph 26);

e) where projects are financed in part by IBRD loans, finance agencies consider cross-default arrangements as a form of insurance (Chapter 9, paragraph 27);

f) member states of the ITU consider setting aside a small proportion of revenues from calls between developing countries and industrialised countries to be devoted to telecommunications in developing countries, for example to fund pre-investment costs (Chapter 9, paragraph 30).

With the longer term in view, WE ALSO RECOMMEND that

g) governments of industrialised countries review their financing instruments and institutions to ensure that they can meet the financing requirements of extending telecommunications networks in developing countries (Chapter 9, paragraph 32);

h) member states of the ITU, in collaboration with international finance agencies, study the proposals for a revolving fund and for telecommunications investment trusts as methods of raising funds for investment in telecommunications with a view to putting these into effect by the next Plenipotentiary Conference at the latest. The Secretary-General is invited to report to the Plenipotentiary Conference on the progress made with these studies (Chapter 9, paragraph 35);

i) the Secretary-General of the ITU, in the light of progress on our other recommendations, study the proposal for an organisation to coordinate the development of telecommunications world wide (WORLD-TEL) and submit his conclusions to the Plenipotentiary Conference (Chapter 9, paragraph 37).

11   Fourthly, to strengthen the role of the ITU, WE RECOMMEND that;

all international organisations concerned with telecommunications give more favourable consideration than hitherto to assistance for the expansion of telecommunications world wide and that regional cooperation be accorded a high priority (Chapter 3, paragraph 13).

12   Finally, WE RECOMMEND that;

the Secretary-General of the ITU monitor the implementation of all the preceding recommendations, report on progress to the Administrative Council and, where necessary, act to stimulate further progress.

13   Our analysis of the problems and the recommendations we have made show that there is no single remedy. A range of actions over a wide front and at different levels is required. Progress will be made only in stages. But, if the effort is sustained, the situation world wide could be transformed in twenty years. All mankind could be brought within easy reach of a telephone by the early part of next century and our objective achieved.

Appendix 2

BT internal management controls, 1987

% Local calls failed due to plant defects
% Local calls failed due to congestion

Exchange connections per employee
Business exchange connection Net Demand
Business exchange connection Net Supply
Business exchange connection Cessations
Residential exchange connection Net Demand
Residential exchange connection Net Supply
Residential exchange connection Cessations

Revenue expenditure
External income
Total Paybill
Gross contribution

Staff in post and % reduction

% Residential orders completed in 8 working days

Public payphones time serviceable
Modern payphone penetration

% PCs completed to target or Customer Latest Date

PC's service restored in 5 working hours
Telex orders completed in 5 working days or Customer Latest Date
Analogue PC faults per circuit per annum

Network faults per exchange connection per annum
% Code 1 service faults cleared by end of next working day

% Repair service calls answered in 25 seconds
Faults per exchange connection per annum

% Trunk calls failed due to plant defects
% Trunk calls failed due to plant engaged

Appendix 3

1. *Telephone service*

    1.1 Size of system
    Telephone stations (sets) of all kinds connected to the public network
    Main lines
    Main lines connected to private branch exchanges (PBX, etc)
    Percentage of main lines connected to automatic exchanges
    Percentage of main lines equipped for direct customer dialling to international destinations
    Percentage of main lines which are residential
    Connection capacity of local public switching exchanges (number of subscriber terminal equipments)
    Trunk and international circuit ends connected to:
    Manual switching exchanges
    Automatic switching exchanges

    1.2 Demand
    New applications for main lines
    Total demand for main lines (including transfers)
    Waiting list for main lines

    1.3 Traffic
    Total national traffic
    Local traffic
    National trunk (toll) traffic
    Total outgoing international traffic
    Total traffic

Outgoing international subscriber dialled traffic as a percentage of total outgoing international traffic

2. *Public telegram service*

   Number of national paid telegrams
   Number of international outgoing full rate telegrams
   Number of international outgoing LT telegrams
   Number of international outgoing phototelegrams

3. *Telex service*

   3.1 Size of telex system: Subscriber lines
   3.2 Traffic: National traffic
                 Outgoing international traffic
                 Total traffic

4. *Data transmission*

   4.1 Size of data system: Number of data terminal equipments on the public telephone and telex networks
   Number of private leased circuits
   Number of data terminal equipments connected to dedicated public data networks

5. *Equivalent full-time telecommunication staff*

   Total staff in telecommunication services
   Operating staff
   Technical staff
   Other staff

6. *Demographic and macro-economic data*

   6.1 Number of inhabitants
   6.2 Number of households
   6.3 Gross domestic product at factor costs in national currency (GDP)
   6.4 Gross fixed capital formation in national currency (GFCF)
   6.5 Exchange rate (national currency equivalent to 1 United States dollar at the end of the year)
   6.6 Consumer price index (1970 = 100)

7. *Income, expenditure and financial results of telecommunication services*

    7.1 Total income from the telephone service
    7.1.1 Income from connection charges
    7.1.2 Income from annual rentals
    7.1.3 Income from calls
    7.2 Total income from the public telegram service
    7.3 Total income from the telex service
    7.4 Other income (data, facsimile transmission services)
    7.5 Total income from all telecommunication services
    7.6 Total current expenditure for all telecommunication services
    7.6.1 Operational expenditure
    7.6.2 Depreciation
    7.6.3 Interest paid
    7.6.4 Taxes
    7.6.5 Other expenditure
    7.7 Income minus expenditure for the telecommunication services

8. *Investments (annual gross construction expenditure) on telecommunication*

    8.1 Total annual gross investments in telecommunications including land and buildings
    8.2 Total annual gross investments in telecommunications excluding land and buildings
    8.3 Annual gross investments for telephone services
    8.4 Annual gross investments in telephone switching equipment

9. *Comparative ratios*

    9.1 Telephone main lines per 100 inhabitants
    9.2 Telephone stations (sets) of all kinds per 100 inhabitants
    9.3 Telecommunications investments as a share of GDP
    9.4 Telecommunications investments as a share of GFCF

Appendix 4

British Telecommunications statistics

The purpose of this appendix is to illustrate, by way of figures, the behaviour of a number of the British Telecommunications variables over the period 1970–88.
Included are the following figures:

Figure A4.1 BT growth in lines (annual growth in exchange connections)
Figure A4.2 BT percentage growth in lines (percentage growth in exchange connections)
Figure A4.3 BT growth in inland calls (annual growth)
Figure A4.4 BT percentage growth in calls (annual growth in inland calls)
Figure A4.5 BT inland calling rate (calls per exchange connection)

A correlation between the operational variables and the fluctuation of the UK economy has been confirmed by statistical analysis.

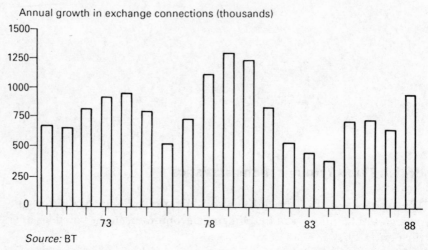

Annual growth in exchange connections (thousands)

Source: BT

Figure A4.1 BT growth in lines (annual growth in exchange connections ('000))

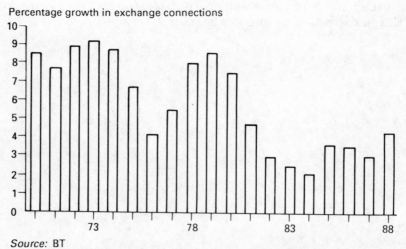

Percentage growth in exchange connections

Source: BT

Figure A4.2 BT percentage growth in lines (percentage growth in exchange connections)

Annual growth in inland calls (millions)

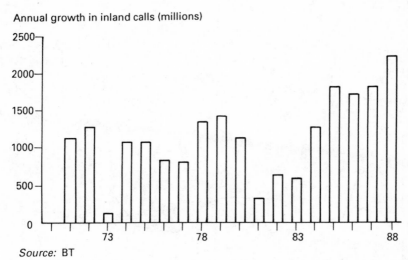

Source: BT

Figure A4.3   BT growth in inland calls (annual growth)

Percentage annual growth in inland calls

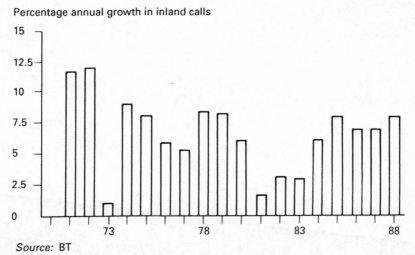

Source: BT

Figure A4.4   BT Percentage growth in calls (annual growth in inland calls)

Calls per exchange connection

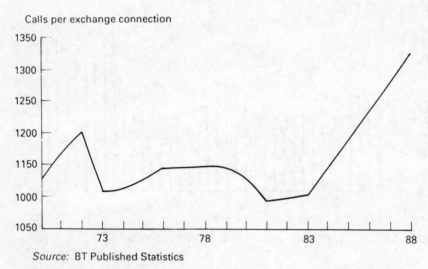

Source: BT Published Statistics

Figure A4.5   BT inland calling rate (calls per exchange connection)

Appendix 5

Evolution of the ISDN

The diagrams that follow illustrate the evolution of the IDN and the ISDN from earlier electro-mechanical networks (see Chapter 10). The stages are:

Step 1 Existing analogue switching; FDM and isolated PCM (digital) transmission systems. This is typical of most older networks round the world.

Step 2 Digital switching and transmission before integration.

Step 3 The Integrated Digital Network (IDN).

Step 4 Digital customer service at 64 kbit/s — elimination of A/D conversion except in telephone instruments. (This configuration may not be met in practice; it is included to illustrate the principle of customer apparatus A/D conversion.)

Step 5 ISDN with customer 2B + D facilities.

Step 5A ISDN on PABXs.

Step 6 Broadband ISDN.

Step 7 Partial integration of CATV facilities.

Step 1 – existing

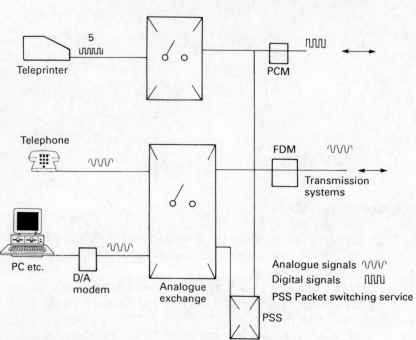

Figure A5.1

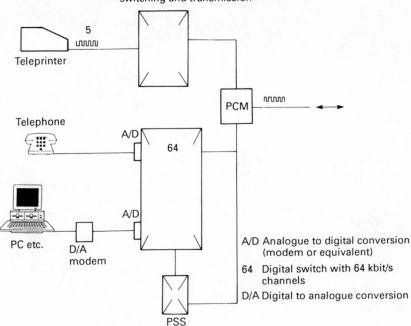

Figure A5.2

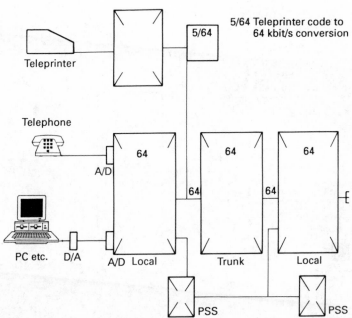

Figure A5.3

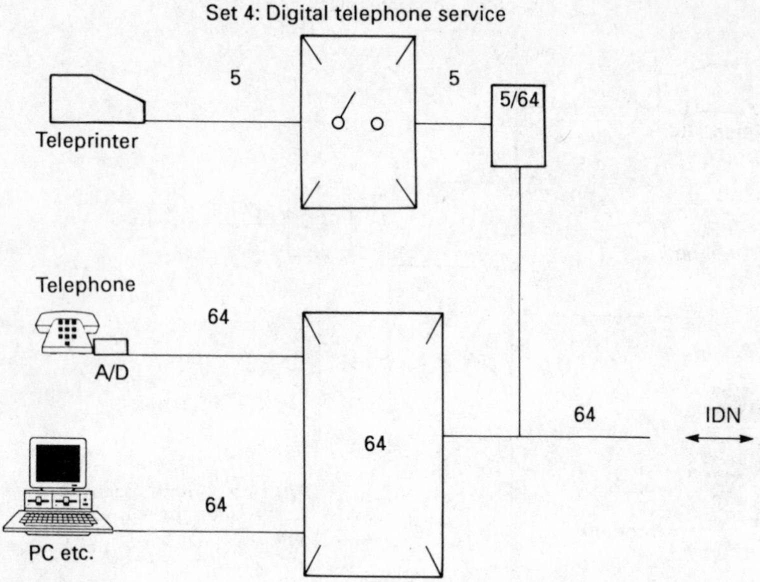

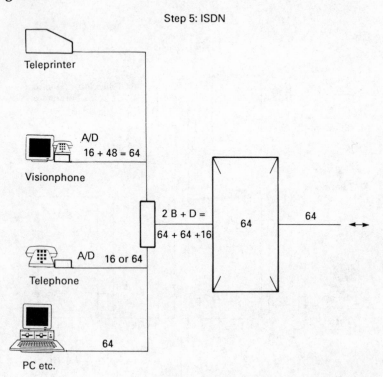

Set 4: Digital telephone service

Teleprinter — 5 — 5 — 5/64

Telephone A/D — 64

PC etc. — 64

64

64 IDN

Figure A5.4

Step 5: ISDN

Teleprinter

A/D
16 + 48 = 64
Visionphone

Telephone A/D 16 or 64

PC etc. 64

2 B + D =
64 + 64 +16

64

64

Figure A5.5

Step 5A: PABX – ISDN

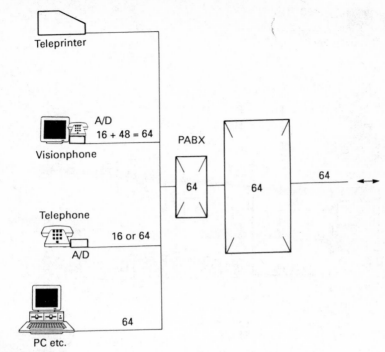

Figure A5.6

Step 6: Broadband ISDN

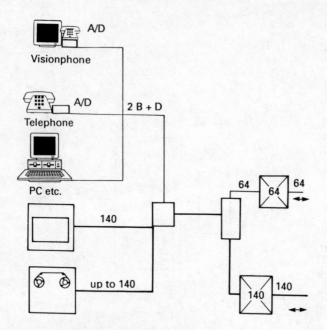

140   140 mbit/s channels
       and switches

Figure A5.7

Step 8 – Add CATV

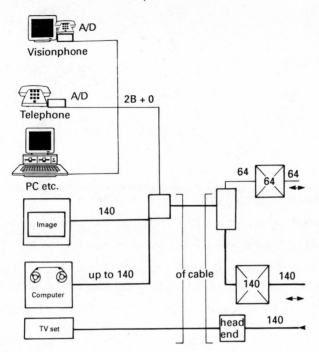

Figure A5.8

Bibliography

The publications referred to by abbreviations below are:

IPOEE Journal — Institution of Post Office Electrical Engineers Journal

BTEJ — British Telecommunications Engineering Journal

A complete set of articles in past editions of these journals is held in the Library of the Institution of Electrical Engineers, Savoy Place, London.

GAS — Studies published by the ITU which are obtainable from International Telecommunications Union General Secretariat, Sales Section, Place des Nations, CH-1211 Geneva 20, Switzerland.

General and economic questions

GAS 5 Economic studies at the national level in the field of telecommunications: New texts for the period 1968–1972.

GAS 5 Economic studies at the national level in the field of telecommunications (1973–1976): 1976 edition.

GAS 5/4 Economic studies at the national level in the field of telecommunications (1977–1980): Special aspects of telecommunications development in isolated and/or underprivileged areas of countries.

GAS 5/9 Economic studies at the national level in the field of telecommunications (1981–1984): Determination of the economic impact of new services on telecommunication undertakings.

Organization and quality practices

Peters, and Waterman, , *In Search of Excellence*, New York, Harper & Row, 1982.

Lomas, , 'Managing for Quality', *BTEJ*, Vol.1 (19..), p.216.

Accounting, Control and Budgetary Systems; and Planning

Littlechild, , *Elements of Telecommunications Economics*, London, Peter Peregrinus, 1979.

GAS 5/5 Economic studies at the national level in the field of telecommunications (1977–1980): Procedure for establishing a budget model for a telecommunication undertaking.

GAS 5/6 Economic studies at the national level in the field of telecommunications (1981–1984): Study of financial and accounting problems related to the effects of inflation on telecommunications authorities.

GAS 5/7 Economic studies at the national level in the field of telecommunications (1981–1984): Study of management information system for telecommunications authorities and appropriate application of the information technology.

GAS 5/1 Economic studies at the national level in the field of telecommunications (1977–1980): Market factors affecting telecommunication demand.

GAS 5/2 Economic studies at the national level in the field of telecommunications (1977–1980): Methods used in long-term forecasting of domestic telecommunications demand and required resources (overall and by main sector).

Charging and Pricing

Information Network System: Telecommunications in the Twenty-first Century Kitahara, Y., London, Heinemann, 1983, esp. pp. 66–7.

Determination of priorities

GAS 5/8 Economic studies at the national level in the field of telecommunications (1981–1984): Optimum allocation and use of scarce resources in order to meet telecommunications needs in urban or rural areas of a country.

GAS 7 Rural telecommunications 1985.

Engineering Efficiency and Manpower Management

A.C. Lord and T. Newby, 'Efficiency in telecommunications motor transport' *IPOEE Journal* (1973/74) **66**, p.32.

Tomlinson, H. 'Productive Labour, IEE Proceedings Vol 130 PtA No1 January 1983.

Merriman, J.H.H. 'Men, circuits and systems in telecommunications', *IPOEE Journal* (1975/76) **68,** p.77.

'Improving underground maintenance', C.C. Thain and D.R. Wells, *IPOEE Journal* (1969/70) **62,** p.222.

Turner, G.E. 'Manpower productivity in telecommunications', *IPOEE Journal* (1971/72) **64,** p.138.

Satellites in Developing Countries

GAS 8 Manual on the economic and technical impact of implementing a regional satellite network, 1983.

Network Planning

Farr, *Telecommunications Traffic, Tarriffs and Costs,* London, Peter Peregrinus, 1988.

H.S. Holmes, 'Telephone traffic recording and forecasting', *IPOEE Journal* (1974/75) **67,** p.201.

C.R.J. Shurrock and E. Davis, 'Inland network strategy', *BTEJ*, Vol. 1 (1982/3), p.143.

J.S. Whyte &c, 'Local exchange renewal strategy', *IPOEE Journal* (1974/75), **67** p.130.

GAS 1 National telephone networks for the automatic service, part C, 1968.

GAS 2 Local telephone networks, 1968.

GAS 2 Local network planning, 1979.

GAS 3 General network planning, 1983.

GAS 3 Economic and technical aspects of the choice of transmission systems, 1986.

GAS 4 Primary sources of energy for the power supply of remote telecommunication systems, 1985.

GAS 6 Handbook on economic and technical aspects of the choice of telephone switching systems, 1981.

GAS 9 Case study on an urban network, 1984.

GAS 9 Economic and technical aspects of the transition from analogue to digital telecommunication networks, 1984.

GAS 9A Case study on the economic and technical aspects of the transition of a complete analogue national network moving to a digital network.

GAS 9B Case study on the economic and technical aspects of the transition of a mixed (analogue/digital) national network moving to a digital network.

GAS 10 Planning Data and forecasting methods 1987 Volumes I & II.

GAS 11 Strategy for the introduction of a Public Data Network in Developing Countries.

VANS and telematics (Chapter 15)

Martin, J., *Telematic Society: A Challenge for Tomorrow* (revised edition), Englewood Cliffs, NJ, Prentice Hall, 1981.

Mayntz, and Schneider, *The Development of Large Technical Systems*, Frankfurt-on-Main, Campus Verlag, 1988, Ch.10.

Nora, S., and Minc, A., *The Computerisation of Society. A Report to the President of France* (English language edition of *L'Informatisation de la Société*), Cambridge, Mass., MIT Press, 1980.

Glossary

2B + D	Digital transmission technique applied to customer distribution circuits providing three channels over one pair of wires (see Chapter 10)
AT&T	American Telephone and Telegraph Company: major United States telecommunications operating company
Analogue principle	Technical principle under which information is conveyed and processed as electrical waves or signals which are the analogue of sound waves, etc.
Area	BPOT field formation (see Chapter 3)
BABT	British Approvals Board for Telecommunications: agency in Britain concerned with testing and approval of CPE (qv) (see Chapter 15)
BPOT	British Post Office Telecommunications: division of the British Post Office concerned with telecommunications; monopoly telecommunications operator to 1981
BT	British Telecommunications plc (British Telecommunications state Corporation from 1981 to 1984): principal British telecommunications operator; transferred to private sector control in 1984 (see Chapter 2)
BTI	British Telecom International: division of BT concerned with international telecommunications
Bildschirmtext	First West German videotext (qv) service
Bit/s	Bits per second: number of 'on-off' bits of information transmitted per second by digital system (see Chapter 9)
CATV	Cable television: technology for distribution of television by physical cables (see Chapters 10 and 11)

CCIR	International Consultative Committee for Radio (the acronym uses the French title): principal ITU organ for radio technology (see also IFRB)
CCITT	International Consultative Committee for Telegraph and Telephone (the acronym uses the French title): Principal ITU telecommunications specialist organ
CCS	Call Connect Systems: generic UK term for PABXs and smaller multi-line customer telephone systems
CPE	Customer Premise Equipment: computing and telecommunications equipment on customer premises
Crossbar	Electro-mechanical exchange system in which connections are made by actuating bars in a crossing matrix
DCF	Discounted Cash Flow: investment appraisal technique in which year-by-year project expenditure and income, typically over twenty year period, is tabulated and discounted using specified interest rate to determine 'present value' or 'internal rate of return' as basis for decision
Digital principle	Technical principle under which information is processed and transmitted in the form of 'on-off' electrical pulses or bits forming binary codes (see Chapter 9)
EFT	Electronic Funds Transfer: application of telecommunications and computing to transfer of funds, e.g. between banks
EFTPOS	Electronic Funds Transfer at Point of Sale: application of telecommunications and computing to transfer funds electronically at retail cash desk (e.g. by credit card reader terminal with on-line connection to central bank computer)
Erlang	Standard measure of telecommunications traffic (see Chapter 12)
GPO	General Post Office: title of British government Department responsible for telecommunications up to 1969
Gallium arsenide	Chemical compound used to make very high speed computing and millimetric radio devices (see chapters 9 and 11)
Gbit/s	Gigabits per second: thousands of millions of bits per second (see Chapter 9)
HDTV	High-definition television: advanced television system delivering pictures of enhanced quality suitable for large screens (see Chapter 10)

IDN	Integrated Digital Network: exchange and inter-exchange network configuration in which all plant in path of a call is digital and digital technical processes are integrated end-to-end (see Chapter 10)
IFRB	International Frequency Registration Board: ITU organ concerned with world-wide registration of transmitter frequencies (see Chapter 13)
INTELSAT	Intergovernmental agency for world-wide satellite communications
ISDN	Integrated Services Digital Network: configuration in which all digital services are delivered to customer over a single physical channel like a pair of wires
ISO	International Standards Organization: principal international standards organization (see Chapter 15)
ITU	International Telecommunication Union: United Nations agency charged with regulation and support of telecommunications (see Chapter 13)
Kbit/s	Kilobits per second: expression used to state number of bits per second tranmitted by digital system (see chapter 9). Kbit/s means thousands of bits per second
KTS	Key Telephone System: UK and US term for simple multi-line customer telephone systems in which extensions select lines by pressing keys
KmH	Kilomanhours: BT specialist expression meaning thousands of manhours
LAN	Local Area Network: computing expression for network on customer premises linking computing equipment (see Chapter 14)
Long Lines	Subsidiary of AT&T (qv) responsible for long distance communications in the USA
MAN	Metropolitan Area Network: computing term for network interlinking computers within metropolitan area like New York or London
MCL	Mercury Telecommunications Ltd: subsidiary of Cable and Wireless plc; second British static telecommunications operator competing with BT
MTP	Medium-Term Plan: coordinated five-year plan for an enterprise (see Chapter 5)
Main line	Basic telephone service connection; a main line is an exclusive, rather than shared, connection to an exchange
Maitland Commission	The Independent International Commission for World-Wide Telecommunications Development. Established by the ITU Plenipotentiary in 1982

Mbit/s	Megabits per second: millions of bits per second (see Chapter 9)
Micro-processor	Processor (qv) fabricated on a single silicon or other semi-conductor chip (see Chapter 9)
Milliseconds	Thousandths of a second
NTT	Nippon Telephone and Telegraph Public Corporation: principal Japanese public telecommunications operator
Non-Voice	Generic term for all telecommunications applications except telephone (speech or voice). Includes data, text, image and vision transmission
OFTEL	Regulatory organ of Government for UK telecommunications (see Chapter 2)
PABX	Private Automatic Branch Exchange: exchange located on customer's premises and owned or leased by him
PC	Personal Computer: desktop-size computer exploiting integrated circuits and built round a micro-processor (qv)
PCM	Pulse Code Modulation: specialist term for digital technology used to encode and convey speech and other information in digital telecommunications systems (see Chapter 9)
PLC	Public Limited Company
PMBX	Private Manual Branch Exchange: operator exchange located on customer premises and owned or leased by him
PSTN	Public Switched Telecommunications Network: principal public telecommunications network used for basic telephone service
Packet-switching	Technique under which information in digital form is assembled in packets of bits for transmission over network (see Chapter 9)
Prestel	First British videotex service (see Chapter 15)
Processor	Central element of computing and modern telecommunications systems which controls all other elements and carries out all basic processes like calculating, sorting and so on
Productiveness	Expression used in this book to denote the overall economic, effectiveness of staff activity (see Chapter 8). (The word 'productivity' is often used with a similar meaning.)
Pulse Code Modulation	see PCM

RMNA	Return on Mean Net Assets: specialist term for form of financial target prescribed by Government for BPOT (see Chapter 7)
RPI	Retail Price Index: the standard UK measure of price inflation
RPOA	Recognized Private Operating Agency: ITU term denoting private sector telecommunications operators like BT recognized by their Governments
Racal Vodaphone Ltd	Subsidiary of RACAL plc: Second cellular radio operator (see Chapter 11) competing with Cellnet (BT/Securicor subsidiary)
Region	BPOT field formation (see Chapter 3)
SMATV	Satellite Master Antenna Television: system for receiving satellite television on central antenna and distributing to customers by cable
SPC	Stored Programme Control: technique under which exchange switching, routing and facility operations are controlled by computer software; Basis of modern digital systems
Silicon	Natural element very widely used as basis of transistors and integrated circuits (see Chapter 9)
Software	Generic industry term for programs held in electronic stores to control behaviour of processors (qv) (see Chapter 9)
Station	An individual telephone or similar terminal or an individual plug outlet. A main line may have any number of stations connected to it through a PABX
Step-by-step	see Strowger
Strowger	Early automatic (electro-mechanical) telephone switching system using selectors which rotate directly under dial or other pulses
TCD	Technical Cooperation Department: department of the ITU concerned with technical advice and support to developing countries (see Chapter 13)
TXE4	BPOT analogue (reed relay) electronic system. TXE4A was a developed version using SPC (qv) control
TXE4A	see TXE4
Telecontrol	Technique for remote control of water,etc valves and similar devices by telecommunications
Telematics	Generic name for public data retrieval and transaction services using public switched networks and directed to mass markets

Telemetering	Technique for remote reading by telecommunications of reservoir gauges, meteorological instruments and similar devices
Tèlètel	First French videotex (qv) service
Telex	Long-established text transmission service using separate network and teleprinter terminals (see Chapter 10)
UICN	Universal Intelligent Communications Network: future concept of telecommunications network wholly controlled by artificial intelligence (see Chapter 9)
UNDP	United Nations Development Programme
VADS	Value Added Data Service: specialist British term for VANS (qv) involving carriage of data
VANS	Value-Added Network Service: service provided over telecommunications links which 'adds value' to customer use of network (see Chapter 15)
Video-conferencing	Technique for interconnecting conference rooms by point-to-point television (see Chapter 15)
Videotex	Generic name for public or private data services using keyboards and computing or TV screens
Vision-phones	Customer terminals which incorporate cameras, screens, microphones and speech receivers to permit combined speech and vision switched service
Voice	Generic term for speech transmission facilities and services — primarily conventional telephone service and traffic
WAN	Wide Area Network: computing term for network interlinking computers over wide area, say, a town, a country or a continent (see Chapter 10)
WP	Word Processor: specialized Personal Computer used to create and edit text

Index

2B + D 53, 97, 100, 105-6
accounting systems 3, 8, 33-35, 39, 63
analogue
    principle 83-84
    transmission 128-9
approval (CPE) 150
Asset Utilization Factors (AUFs) 127

bilateral loans 64
Bildschirmtext 157-8
bills 57
binary code 83-85, 88, 90-91
bit rates 85
broadband 97, 100, 103, 106
buildings 27-8, 128-9

cable
    comparison with radio 116-9
    distribution planning 128
    installation staff 75
    optical fibre, 110, 128, 138-9
    submarine 138-9
cable television (CATV) 8, 102-4, 106
call(s) 36-7, 82, 90
    charges 31, 49-52, 55-8
    conference 153
    duration 36, 124
    failure 36-7, 126
    forecast 124-6
    long distance 51, 53-4, 55, 90, 125
    service function 16, 36
    routing 100-1, 139
    sampling 37
    timing 50
call connect systems (CCSs) 114, 119, 145,
    150
capital 9, 58-65, 129-30, 170-1
    return on 51-2, 63
cardphones 147
cashflow 122
CCIR 133-7
CCITT 127, 133-7, 155

cellular radio systems 3, 8, 12, 111-3, 115-9
Centre for Telecommunications
    Development (CTD) 3, 70, 136, 176-7
chargehands 75
charges 3, 9, 49, 53-8, 171
    call 31, 36, 49-52, 55-8
    connection 52-7, 62-4
    future development 53-5
    local call 55
    presentation of 57
    principles 57
Circuit Switched Data Networks 94, 96
Closed User Groups (CUGs) 156-7
competition 8, 11-3, 169
    in CPE 13, 149-152
    in procurement 148
compression of TV signals 163-4
computer
    based training 71
    billing 55
    directory enquiries 28-9
    hardware 85-7
    model, BPOT network policy 121
    networking 100, 146
    programming 88-9
    traffic 105, 138
computers 34, 43, 81-3, 90
    lap-top 87
    mainframe 86, 145
    mini 86
    personal 86-7, 144-5
computing 3, 27, 153-4, 171
    staff 66
    technology 81-9
conference calls 153
Confravision 163-4
congestion 101, 127
connection charges 52-7, 62-4
constitutional developments
    in Britain 7-8, 169
    in CPE 169
control systems 35-9

cordless CPE 113-4, 119
credit card verification 159
Customer Premise Equipment (CPE) 64,
   75, 92, 96, 117, 119, 142-152, 172
   approval 150
   cordless 113-4, 119
   divestment of 3, 8, 13, 27, 151
   managerial & policy problems 3, 8, 12-3,
      27, 147-9
   procurement 147
   testing 149-50

data 91, 94, 100, 106, 139, 145-6
databases 101, 154
demand for service 36
   forecast 117, 123
   management 52-3
depreciation 63
design
   buildings 27-8
   of equipment 131
   periods 127
digital
   codes 83
   exchanges 71, 84
   microwave 109-10
   network 83-5, 96-100, 137
   PABXs 145-6
   principles 83-5
   systems 83, 127
   transmission 90-2, 127
dimensioning 122-4
Direct Broadcasting by Satellite (DBS) 104
directories 16, 28
directory enquiries 28, 32, 37-47,
   on computer 28, 157
Discounted Cash Flow (DCF) 122
distance-related charges 53-4
duct 117, 128

earth stations 138
Electronic Funds Transfer at Point of Sale
   (EFTPOS) 157-60
electronic mail 161
Electronic Funds Transfer (EFT) 158-60
engineering managers 75-6
engineering staff 9, 24, 26-9, 32, 37, 71-2,
   75-6, 148
exchanges 25, 49, 83, 90, 101, 121-3, 126-7
   analogue 84
   capital cost 121
   digital 71, 84
   electro-mechanical 71, 93-4
   forecasts 124-6
   international 101, 139-40
   layout 123
   local 95-7, 101
   maintenance of 27, 71
   telex 93
expedients – short term 129
externally generated funds 64-65

facsimile 142, 144, 145, 152
faults 74
field
   general managers 17-22, 44-5
   structures 15-22
financial sector services 157-60
financial objectives 42-3, 50-2
first line managers xiii, 72, 75
five year 43, 45-7, 60
flat rate 49-50
forecasts 45-6, 121, 123
   building 128-9
   central planning 41
   error 127
   in plant planning 124-6
frequenc(ies) 108-111, 113, 137
   bandwidth 84
   spectrum, lack of 112
Frequency Division Multiplexing (FDM) 90

gallium arsenide 85
General Managers – field 17-22, 44-5
Government 7-9, 12, 33, 42, 52, 59, 63, 132,
   173
grade of service standards 126
grading structure of staff 75

handsets 117-8
hardware 85-6, 155
High Definition Television (HDTV) 103,
   107

INTELSAT 138
incentives – staff 74
Independent International Commission
   xiv, 1, 58-9, 62, 65, 71, 136, 174-8
industrial relations 78
installation controls (CPE) 148
IDN 53-5, 95-6, 100, 105-6, 160-1
ISDN 53-5, 96-100, 106, 118
intelligent network 101-2
interfacing 155-7, 165
IFRB 109, 134-5
ISO 155
ITU 2-3, 30, 132-136, 172, 176-8
inventory control 148
investment 9, 58-60
   requirements 60-2, 65

key telephone systems (KTS) 145, 152
keyboards 81

laboratories, test 150
leased circuits *see* private circuits
liberalization (CPE) 149-52
line(s)
   demand for 123
   forecast 123
   plant planning 128
   shortage 122
   surplus 122
loans 63

Local Area Networks (LANS) 146
local loop 12, 24, 91
long distance calls 125
Long Lines 25-6, 32

mailboxes 145, 153, 156
maintenance 27
  costs 121
  CPE 148, 151-2
  manhours 74
  staff 127
management
  of staff 66-7
  staff 70
  training 70-1
managers 127, 149
  engineering 75-6
  first line 75
  operating 86
manufacturing industry 132
  role in training 176
Mercury Communications Ltd (MCL) 8, 35,
  132, 134
  payphones 147
meters 55, 57
Metropolitan Area Networks (MANs) 100
micro-processors 86-7, 113
micro-cellular 116, 119
microwave
  competition with cable 110
  frequencies 109
  stations staff 75
  systems private 105
millimetric frequencies 108, 111
Millimetric Microwave Video Systems
  (MVDS) 106
Minitel (Teletel) 157-8
mobile radio 111-2, 117
  private & public 111-2
  terminals 111
mobile exchange equipment 129
modems 96
motor transport staff 75
multiplexing 55, 101, 123

National Networks Division (BT) 25
network(s) 2, 12, 122, 171
  basics 100-1
  capacity 123
  circuit switched 96
  dimensioning 123
  domestic 139
  economics of 127
  expansion 1 60
  financial sector 157-160
  intelligent 101
  international 139
  links 116
  local area (LANs) 12, 25, 146
  non-voice 53, 171
  operators 150
  planning 121-131, 170-2

plant procurement 129-31, 171-2
  private 25, 104-106
  processor controlled 55-6, 91, 93, 120-3,
    127, 136
  PSTN 93, 106, 120-31, 153-165
  replacement policy 66
  self-adapting 156
  services 151, 153-165
  service performance 127
  structure 101
Network Control Centre 101
networking, computer 100, 146
nodes 123
non-voice
  cordless CPE 113-4
  services 170
  terminals 140, 155
Nora/Minc Report 161

Office of Telecommunications (OFTEL) 8
one year plan 41, 43-45
one-per-desk 145
Open Systems Interconnect (OSI) 155-6
operator(s) 8, 12
  services 57, 76-7
  training 77
optical fibres 91-2, 110, 128
  submarine cable 138
  TV systems 103-4
organization 9, 22, 30-2, 169
  internal 3, 9, 26-7
  new products & services 2, 28
  structures 15-23, 38-9, 44-5
  tiers of 23
  workforce 71-2, 75-6
overhead wire 128
overtime 73
ownership 2
  changes in 8

packet switching 54, 139-40
  international 140
  network 96, 120
payphones 147
penetration 60-61
personal communicators 116, 117
personal computers (PCs)
  on cellular 113
  communicating 104, 142
  fax boards 144-5
  Prestel 156-7
personal disposable income 41
personnel
  planning 78
  selection 70
picture transmission 139
planning 3, 9, 24, 60-2, 121-131, 170, 176
  building 128-9
  distribution cable 128
  local cable network plant 122, 128, 131
  local line 128
  personnel 78

plans
    five year 43, 45-7, 60
    one year 41, 45-8
plant 59, 90-9
    costs 50
    international 92
    layout 101
    manual 123
    network 121-37
    planning 92, 123
    shortage 62, 131
    surplus 131
plugs and sockets (CPE) 143-4
pole(s) 128
    erection gangs 75
portable terminals 113
Post Office, The 8
posts 10
power costs 121
precision of charging 55-6
Prestel 156-7
pric(ing)
    elasticity 61-2, 57
    principles of 3, 9, 50-2, 170
printers 81
priorities 58-60, 170-1
PABXs 96-7, 105-11, 145-6, 149, 151-2, 155
    analogue 145
    cordless 119
    digital 145
    maintenance 75
private circuits 104-7, 118-9, 159
PMBXs 145, 152
private networks 25, 140, 159
private sector 8
    funding 65
privatization 8, 12
processors 82, 91, 117
procurement 128, 172, 176, 129-31
    CPE 148-151
product groups 27, 29-30, 32
productiveness 71-4, 78, 201
profit centres 38, 74
project
    appraisal 121-2
    forecasts 124-6
protocol conversion 155-6
public call offices 36
Pulse Code Modulation (PCM) 90

quality 23
    assurance 36-9, 74, 78, 130-1
    of service 126

"RPI-x" 51
radio 3, 92, 108-19, 171-2
    mobile 111-3, 117-8
    paging 115
    private 118-9
regions (in BPOT) 19, 44
rental 49-50, 63-4
replacement policy 40, 42, 63, 93, 120-2

retrieval (database) services 101, 154
return on capital 42, 50
Return on Mean Net Assets (RMNA) 63
rout(ing)
    duct 128
    pole 128
    within networks 96, 122-4, 127
rural telecommunications 115-6, 176

satellites 3, 104, 106, 110-1, 118, 138-5, 164
SMATV 104, 106
self-certification 150
self financing ratio (SFR) 63
separation
    from Government 2, 8, 10-11, 13, 169
    from Posts 10-11
services
    international 3, 92
    network 151, 153-165, 173
    new 2, 29
    value added *see* VANS
    visual 153, 162-4
software 87-9, 117, 155
spectrum 108-11, 113, 117-9, 134
speeds (bit rates) 85, 91, 97, 163-4
staff 3, 8, 40, 59, 66-78, 116, 131, 171
    administrative & office 37, 69, 73, 75, 77
    computing 69
    engineering 9, 24, 26-9, 32, 37, 71-2, 75-6,
        148
    managers 70, 77, 127-8
    operating 26, 28, 68, 73, 76-7
    recruitment 40, 50, 70
    requirements 43, 66-70, 75, 78, 127
    supervisory 76-7
    traffic 16
    training 40, 59, 70-1, 77
    wastage 70
    women 73, 77
standards 136, 145, 154-5
statistical systems 3, 33, 35-39
statistics 36-7, 179-182
SPC
    exchanges 55-6, 91, 93, 120-3, 127, 136
    PABXs 145-6
submarine cables 109, 137-9
supervision 73
suppliers 124, 129-31
    delivery performance 127
switched star CATV systems 102

tariffs 50
taxation 9
technical authority 131
telecontrol 164-5
telegrams 137
telematics 53, 156-62, 165
telemetering 164-5
telephone(s) 7, 36, 74, 81, 142-3, 148, 152
    answering service 153
    charges 49-57, 137
    cordless 113-4, 119

purchase by customer 143
  supply 148-51
Telepoint 114
teleprinters 142, 144, 152
Teletel 157-8, 161-2
Teletex 156
television sets as terminals 156-7
telex 36, 96, 145, 156
  charges 54
  international 139
  network 93-4, 105, 120
tendering 130
terminals 145, 152-3, 155-7
  data 95
  international 101
  portable 113-5
  satellite 101
text services 156-8
traffic
  capacity 122
  forecasts 124
  measurement 101, 127
  non-voice 54-5, 140
  routing 101, 123
  voice 54, 56, 138, 140, 153

training 59, 70-1, 78, 176
transaction services 102
transmission 101
  losses 101
  plant 122-3, 126
  speeds 54, 95
transistor(s) 85
tree and branch 103

UICN 102, 156

VADS 3, 92, 153
VANS 3, 12-3, 29, 55, 151, 153-165, 173
valve(s) 85, 90, 138
Very Large Scale Integration (VLSI) 85
video conferencing 162-4
Videotex 153, 156-7, 162-4
Viewdata 153
vision phone 105, 164
visual services 153, 162-4

waiting lists 42, 52, 62, 128
Wide Area Networks (WANS) 100, 105
word processors (WP) 87, 153

X 400 144, 155-6